SEASHELLS
of Georgia and the Carolinas

A Beachcomber's Guide

SEASHELLS
of Georgia and the Carolinas
A Beachcomber's Guide

Blair and Dawn Witherington

Pineapple Press, Inc.
Sarasota, Florida

To our parents

Front Cover Photographs

Turkey wing (zebra ark) *(Arca zebra)*
Lion's-paw *(Nodipecten nodosus)*
Banded tulip *(Fasciolaria lilium)*
Knobbed whelk *(Busycon carica)*
Variable coquina clam *(Donax variabilis)*
Clench (cameo) helmet *(Cassis madagascariensis)*
Common nutmeg *(Cancellaria reticulata)*
Scorched mussel *(Brachidontes exustus)*
Atlantic giant cockle *(Dinocardium robustum)*

Back Cover

Florida rocksnail *(Stramonita haemastoma)* on an Oregon Inlet jetty, NC

Front Flap

Paper nautilus, eggcase of the greater argonaut *(Argonauta argo)*

Inquiries should be addressed to:

Pineapple Press, Inc.
P.O. Box 3889
Sarasota, Florida 34230

www.pineapplepress.com

Library of Congress Cataloging-in-Publication Data

Witherington, Blair E., 1962-
  Seashells of Georgia and the Carolinas : a beachcomber's guide / Blair and Dawn Witherington.
— 1st ed.
    p. cm.
  Includes bibliographical references and index.
  ISBN 978-1-56164-497-1 (pb : alk. paper) 1. Shells—Georgia—Identification. 2. Shells—
North Carolina—Identification. 3. Shells—South Carolina—Identification. 4.
Mollusks—Georgia. 5. Mollusks—North Carolina. 6. Mollusks—South Carolina. I.
Witherington, Dawn. II. Title.

  QL415.GW58 2011
  594.147'709758—dc22

First Edition

2010053191

Design by Blair and Dawn Witherington
Printed in India

Contents

Acknowledgments

For their contributions, review, and advice we are indebted to Dean Bagley, Carly DeMay, Bill Frank, Kim Mohlenhoff, Chaz Wilkins, and the Outer Banks Beachcomber Museum—home of the Nellie Myrtle Pridgen collection.

Back flap author photo is by Dean Bagley and middle image p. 35 is by Carly DeMay. All other images are © Blair Witherington and Dawn Witherington.

Guide Organization

Each seashell in this book has a **range map** showing where one might find it. These ranges pertain to an item's beach distribution, which may be different from the places where the mollusk lives. Coastal lines on the maps are solid where an item is relatively common, and open where relatively uncommon. Because the range maps are not absolute, a gap may indicate either rarity or uncertainty. The shell **sizes** given in each shell's description refer to maximum length unless otherwise indicated.

Relatively Common   Relatively Uncommon

Because this is a guide to beach-found seashells, the depictions that follow are of beached shells. That is, many are likely to show the characteristic wear resulting from surf tumbling. This beach wear often creates forms that look different from the museum specimens portrayed in most shell books. Although most shells on beaches are dead, many mollusks live their lives in the surf zone. These living animals leave many interesting signs of life other than their shells.

Introduction

Seashells on beaches satisfy the searcher in each of us. They attract us with all it takes to trigger the collection compulsion—beauty, variety, mystery, intrigue, and a pocket-sized form. Seashells also happen to adorn one of our favorite places—the beach, where a barefoot, sandy-seaside stroll can reveal a bounty of collectibles even to a casual visitor. But for searchers with their mind's eye sharpened by images of potential finds, there is a diverse world spiced with rare and provocative shells that most others would simply pass by.

So what are these seashells? They are, or were, the protective and supportive parts of soft-bodied marine mollusks—animals in the phylum Mollusca. For most shells found on beaches, only the animal's persistent parts remain after its softer bits have fed other elements of the food chain. Shells persist on beaches due to their mineral makeup—calcium carbonate crystals laid down in opposing directional layers held together by small amounts of protein glue. Of course, this chemical description does nothing to explain the persistence of seashells within the long history of the human experience. Only the natural poetry of a shell's exquisite form can do that.

Indeed, seashells are striking artistic works of nature, with no two quite the same. Many collectors are fascinated with this diversity even within a single species and may amass parochial collections showcasing the variation on a common theme. Other shell collectors may wish to represent all the compelling forms that our area's mollusk fauna have to offer.

In this book, we represent four classes of mollusks that leave behind collectible shells and that occasionally make a living appearance on Georgia and Carolina beaches.

Gastropods (class Gastropoda, meaning "stomach-footed") are symmetrical animals twisted into an asymmetrical shape. These are the familiar spirally coiled snails, but this class also includes cone-shaped limpets and the sea slugs, which have internal shells or no shell at all.

Scaphopods (class Scaphopoda, meaning "boat-footed") are eyeless animals that live in tusklike shells that are open at each end.

Cephalopods (class Cephalopoda, meaning "head-footed") have distinct heads, complex eyes, and a set of head-appendages (tentacles) that surround the mouth. For most, including octopodes, cuttlefish, and squid, the shell is internal.

Bivalves (class Bivalvia, meaning "two-shelled") have valves (shells) connected by a hinge and include the familiar oysters, scallops, and clams. This class also includes shipworms, which have reduced shells and live within sea-soaked wood.

Finding Beach Shells

Which beaches have the best shelling? A few beaches do stand out, and a list of these is presented with the map below. It's a long list because no short list could fully represent all the seashells that Georgia and the Carolinas have to offer. Shells that are common on some beaches are rare on others. Beachcombers who seek a diverse collection,

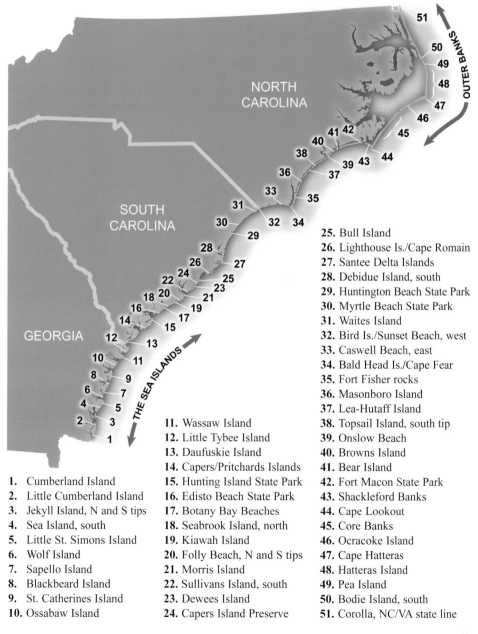

25. Bull Island
26. Lighthouse Is./Cape Romain
27. Santee Delta Islands
28. Debidue Island, south
29. Huntington Beach State Park
30. Myrtle Beach State Park
31. Waites Island
32. Bird Is./Sunset Beach, west
33. Caswell Beach, east
34. Bald Head Is./Cape Fear
35. Fort Fisher rocks
36. Masonboro Island
37. Lea-Hutaff Island
38. Topsail Island, south tip
39. Onslow Beach
40. Browns Island
41. Bear Island
42. Fort Macon State Park
43. Shackleford Banks
44. Cape Lookout
45. Core Banks
46. Ocracoke Island
47. Cape Hatteras
48. Hatteras Island
49. Pea Island
50. Bodie Island, south
51. Corolla, NC/VA state line

11. Wassaw Island
12. Little Tybee Island
13. Daufuskie Island
14. Capers/Pritchards Islands
15. Hunting Island State Park
16. Edisto Beach State Park
17. Botany Bay Beaches
18. Seabrook Island, north
19. Kiawah Island
20. Folly Beach, N and S tips
21. Morris Island
22. Sullivans Island, south
23. Dewees Island
24. Capers Island Preserve

1. Cumberland Island
2. Little Cumberland Island
3. Jekyll Island, N and S tips
4. Sea Island, south
5. Little St. Simons Island
6. Wolf Island
7. Sapello Island
8. Blackbeard Island
9. St. Catherines Island
10. Ossabaw Island

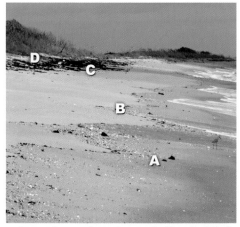

Lettered areas show where to find beached shells

A patch of small shells assembled on the low beach

Seashells often cluster in predictable patterns

or merely a diverse experience, will want to visit a wide array of shelling destinations. But within this list are the particularly shelly beaches downdrift (south) of the region's major shoals, those off of Cape Hatteras, Cape Lookout, Cape Fear, and Cape Romain. Add to these shell sources the many tidal shoals off Georgia's Sea Islands.

To more finely target your shelling, the most important factor determining where to look is the specific search area on the beach you've chosen. These specific hot spots include stretches of beach near inlets, at the tips of islands, and near offshore reefs.

Knowing a little beach anatomy also helps. Where to look along a beach's width (high or low) depends upon the kind of shells you seek and how the ocean has recently treated the beach. Combing the wet sand at low tide (**A**) is the best way to find small and delicate seashells, especially when the surf is calm. Calm seas may allow shellers to search the surf itself, including the highly productive drop-off (step) just seaward from the low-tide line on some beaches. The recent high-tide line at mid-beach (**B**) is often a good place to find large or fluttery shells. Keep in mind that the high-tide mark, the wrack line (**C**), from previous days may have been higher up the beach where shells can be found if they haven't been covered with wind-blown sand. The largest waves during the highest tides sweep up the beach to the storm wrack (**D**), which is often at the base of the dune but occasionally extends into areas of overwash behind the dune. This area can hold large shells, and on infrequently combed beaches this upper beach could be filled with rare finds.

When is the best time to go shelling? Why, every chance you get, of course. But folks with busy lives may appreciate some explicit guidance. Generally, winter and spring seasons have the greatest abundance of beached shells. These are the Atlantic Ocean's tumultuous months when deep swells move offshore seashells close to beaches. Cool-month nor'easter storms that last more than a day bring great shelling, but the tropical storms and hurricanes of summer and fall also churn up splendid shell diversity including some shells rarely seen at any other time. Don't fret; there's no need to brave a storm's fury in order to reap from its tide of seashells. The best seashell abundance and diversity often occurs on beaches during the calm days immediately following this rough weather.

On almost any day of the year, the best shelling is at low tide. Searches are especially successful and interesting near the full- and new-moon periods of the month, which mark semimonthly spring tides when low tides are lowest. These periods give shellers a unique opportunity to explore the sea bottom without holding their breath. In addition to shells, the receding ocean also reveals many of the living mollusks that make the surf their home. Among the signs these animals leave are characteristic tracks, burrows, egg masses, and occasional whole animals awaiting the sea's return. Many of these living low-tide shells are animated by their second (or third, fourth...) owners, the hermit crabs. Take care to inspect these shells before declaring them dead. Mistakes can lead to ecological guilt and a very stinky collection bag.

A thinstripe hermit crab peers out from a knobbed whelk shell

Shelled Mollusk Anatomy

Seashells are the skeletons of mollusks (phylum Mollusca). The most common shells are from snails (with one shell, most often coiled) and **bivalves** (with two hinged shells). Snails are **gastropods**, as are sea slugs, which have an internal shell or none at all. Other shelled mollusks include tusk shells and some squids.

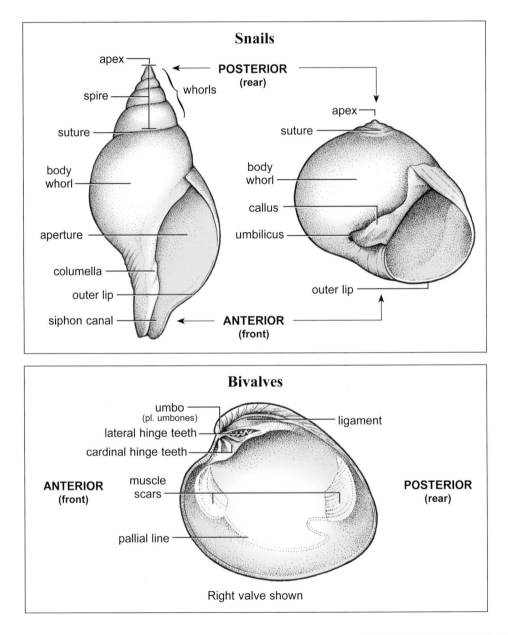

Snails

apex
POSTERIOR
(rear)
spire
whorls
apex
suture
suture
body whorl
body whorl
callus
aperture
umbilicus
columella
outer lip
outer lip
siphon canal
ANTERIOR
(front)

Bivalves

umbo
(pl. umbones)
ligament
lateral hinge teeth
cardinal hinge teeth
ANTERIOR
(front)
muscle scars
POSTERIOR
(rear)
pallial line
Right valve shown

Limpet, Falselimpet, and **Wormsnail**

Keyhole Limpet   Falselimpet   Wormsnail

Cayenne keyhole limpet, max 2 in (5 cm)

RELATIVES: Keyhole limpets (family Fissurellidae) are not directly related to falselimpets (Siphonariidae). Wormsnails are in the family Turritellidae.

IDENTIFYING FEATURES:

Cayenne keyhole limpets *(Diodora cayenensis)* are shaped like oval-based volcanoes with a keyhole opening offset from center.

Striped falselimpets *(Siphonaria pectinata)* have a limpetlike shell with radiating brown stripes and no top opening.

Striped falselimpet, max 1 in (2.5 cm)

Florida wormsnails *(Vermicularia knorrii)* grow wormlike after reaching about 0.5 in (1.25 cm), but also may grow in masses of many intertwined snails. These snails are generally brown with a white spiral tip.

HABITAT: Limpets and falselimpets live on rocks. Limpets are in shallow water, and striped falselimpets are in and above the wave splash. Florida wormsnails live in sponges near shore out to 360 ft (110 m).

Florida wormsnail, max 3 in (8 cm)

DID YOU KNOW? Limpets and falselimpets feed on algae using their rasping, tonguelike radula. Limpets anticipate low tide and return to resting spots before the water leaves their rock. Their top hole passes exhaust from their gills. Falselimpets are pulmonate snails (they have lungs and breathe air) and can be found above the tide on jetty rocks.

A knotted mass of Florida wormsnails

Sculptured topsnail, max 1 in (2.5 cm)

Chestnut turban, max 2 in (5 cm)

Common sundial, max 2.5 in (6.4 cm)

Topsnail, Turban, and Sundial

Topsnail   Turban and Sundial

RELATIVES: Topsnails are in the family Trochidae. Turban shells are in the family Turbinidae. Sundials are in the family Architectonicidae.

IDENTIFYING FEATURES:

Sculptured topsnails *(Calliostoma euglyptum)* have a dark-tipped apex and rounded whorls with beaded cords.

Chestnut turbans *(Turbo castanea)* have rounded whorls like a knobby turban and have a circular aperture. They are occasionally bright orange.

Common sundials *(Architectonica nobilis)* have a deep umbilicus and a spire like a flattened cone.

HABITAT: Topsnails live on nearshore hard bottom and offshore reefs. Turbans and sundials live in sandy areas out to moderate depths.

DID YOU KNOW? Topsnails feed on algae, detritus, small bottom animals, and sponges. An extensive commercial harvest of large, Indo-Pacific turban shells fuels a high-end market for shell buttons. The snails are also the typical substitute "escargots" for the region's tourists. Our Southeastern turbans are too small for this enterprise, but are occasionally eaten locally. Most chestnut turbans feed on algae. Sundials spend their days buried spire-down in the sand and emerge at night to feed on sea pansies.

Wentletraps

Many-ribbed wentletrap, max 0.5 in (1.3 cm)

RELATIVES: Wentletraps are in the family Epitoniidae.

IDENTIFYING FEATURES:

Many-ribbed wentletraps *(Epitonium multistriatum)* are sharply pointed with about 19 crowded ribs on the body whorl, the last of 7–8 whorls.

Angulate wentletraps *(Epitonium angulatum),* have a rounded, thick-lipped aperture and distinct, widely spaced ribs. The body whorl has 9–10 thin ribs that are angled at the whorl shoulders.

Angulate wentletrap, max 1 in (2.5 cm)

Humphrey's wentletraps *(Epitonium humphreysii)* have 8–9 thick, rounded ribs on the body whorl.

Brown-band wentletraps *(Epitonium rupicola)* have rounded ribs of varying strengths and spiral bands of white, tan, and brown.

HABITAT: Wentletraps live in sandy areas to moderate depths. Wentletrap shells vacated by the mollusk are likely to be occupied by long-wristed hermit crabs.

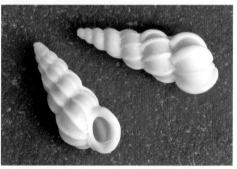

Humphrey's wentletrap, max 1 in (2.5 cm)

DID YOU KNOW? Angulate wentletraps get away with chewing chunks off living anemones by soothing them with a purple anesthetic. More than 20 wentletrap species are known from the southeastern US. "Wentletrap" comes from *wendeltreppe,* German for a winding staircase.

Brown-band wentletrap, max 0.75 in (2 cm)

Marsh periwinkle, max 1 in (2.5 cm)

Zebra periwinkle, max 0.5 in (1.3 cm)

Live zebra periwinkles on a granite rock jetty

Periwinkles

RELATIVES: Periwinkles are in the family Littorinidae.

IDENTIFYING FEATURES:

Marsh periwinkles *(Littoraria irrorata)* have thick aperture lips and are patterned with dashed streaks on their spiral ridges.

Zebra periwinkles *(Echinolittorina placida,* or *Littorina ziczac)* have white and purple-brown wavy lines. Some have a dark middle band (**A**).

HABITAT: Periwinkles live at the high-tide line on firmly anchored substrates. Marsh periwinkles live almost exclusively on marsh grass, and zebra periwinkles live on rocks, especially jetties, but also live on fallen boneyard-beach oaks.

DID YOU KNOW? Periwinkles forage only at high tide and avoid both terrestrial and marine predators by splitting the difference between the two realms. Marsh periwinkles have a unique farming relationship with the fungi they eat. The snail chews wounds into live marsh grass, which becomes infected with fungi. Marsh periwinkles return frequently to graze in their fungus farm, moving up and down the grass stem with each tidal cycle. A century ago, zebra periwinkles were only on rare rocky beach outcroppings in the Gulf of Mexico. But following widespread installation of jetties, the species hop-scotched into the Atlantic as far north as North Carolina.

Ceriths and Eulimas

Dark Ceriths, Niso, and Melanella

Fly-speck Cerith

Dark cerith, max 1.5 in (3.5 cm)

RELATIVES: Dark and fly-speck ceriths (family Cerithiidae) are distantly related to nisos and melanellas (eulimas in the family Eulimidae).

IDENTIFYING FEATURES: Ceriths are lumpy with distinct siphon canals opposite their pointed spires. Eulimas are glossy smooth cones.

Dark ceriths *(Cerithium atratum)* have 18–20 beaded ridges per whorl and occasional larger lumps. Beached shells vary from light to dark.

Fly-speck cerith, max 1 in (2.5 cm)

Fly-speck ceriths *(Cerithium muscarum)* have 9–11 ridges per whorl that are crossed by spiral lines. New shells are "fly-specked" with rows of dots.

Brown-line nisos *(Niso aeglees)* have an angled body whorl and a thin brown line on each whorl suture.

Intermediate melanellas *(Melanella intermedia)* are glossy white with a rounded body whorl.

Brown-line niso, max 0.8 in (2.1 cm)

HABITAT: Dark ceriths live in shell rubble. Fly-speck ceriths live in seagrass. Eulimas live on sea cucumbers or other echinoderms within their varied habitats. All are in shallow waters less than 100 ft (30 m).

DID YOU KNOW? Ceriths feed on algae and detritus. Eulimas feed by chewing through the skin of sea cucumbers or urchins and sucking their blood.

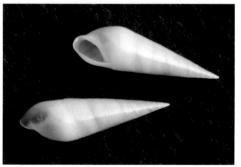

Intermediate melanella, max 0.5 in (1.3 cm)

7

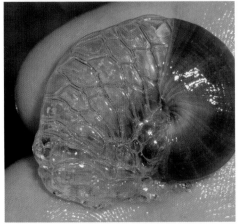

Common purple sea snail, max 1.5 in (3.5 cm)

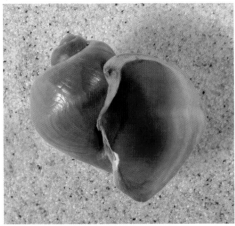

Common purple sea snail with bubble raft

Janthina (Bubble-raft) Snails

RELATIVES: Janthina snails are in the family Janthinidae, distantly related to wentletraps.

IDENTIFYING FEATURES: These gastropods have fragile shells. Live snails have an elastic, translucent, bubble raft arcing from their aperture.

Common purple sea snails *(Janthina janthina)* have a low spire and D-shaped aperture. Their top whorls are pale and their base is violet.

Elongate janthina *(Janthina globosa)* are violet all over with rounded whorls, indented sutures, a distinct spire, and a pointed lower aperture.

HABITAT: Janthina snails live adrift on the open ocean. Unbroken snails are found in freshly beached wrack.

DID YOU KNOW? Two rarer species are known from Southeastern beaches. Pale janthina *(Janthina pallida)* are pale purple with a round aperture. Brown janthina *(Recluzia rollandiana)* are brown with a round aperture and sharp spire. The bubble raft of the brown janthina is also brown, to match the floating sargassum where it lives. Janthina snails cannot swim and will sink into oblivion if they lose their raft. They often sail along with, and prey upon, Portuguese men-o-war and by-the-wind sailors. The snails' violet color blends with the color of the deep ocean and hides them from the birds and young sea turtles that eat them.

Elongate janthina, max 0.8 in (2 cm)

Conchs

Fighting Conch

Milk Conch

RELATIVES: These are true conchs (pronounced "konks") in the family Strombidae.

IDENTIFYING FEATURES:

Florida fighting conchs *(Strombus alatus)* are thick-shelled with blunt-knobbed whorls. Colors vary from pale yellow to chestnut-brown with occasional light spots and zigzags. The body whorl has fine spiral cords (ridges).

Florida fighting conch, max 4 in (10 cm)

Milk conchs *(Strombus costatus)* are similar to Florida fighting conchs but have a more pointed spire and a more widely flaring aperture lip in adults. Their color is milky white.

HABITAT: Both of these conchs live offshore on reefs out to 120 ft (37 m) deep, especially the reefs south of Cape Lookout, NC.

DID YOU KNOW? Fighting conchs get their name from occasional bouts between rival males. They are spry for snails and can quickly flip themselves and walk using their pointed operculum. Both species feed on algae and detritus. Fighting conchs are being farmed experimentally as an edible alternative to the rarer and slower-maturing queen conch *(Strombus gigas)* of the Caribbean. Milk conchs are more common in the Bahamas and southern Florida. Although rare this far north, they occasionally wash in between Cape Fear and Cape Lookout following hurricanes.

Juvenile Florida fighting conchs

Milk conch, max 8 in (23.3 cm)

9

Spotted slippersnail, max 1 in (2.5 cm)

Atlantic slippersnail, max 2.5 in (6.5 cm)

An arching stack of Atlantic slippersnails

Slippersnails *(Spotted and Atlantic)*

Spotted Slippersnail

Atlantic Slippersnail

RELATIVES: Slippersnails share the family Calyptraeidae with cup-and-saucer snails.

IDENTIFYING FEATURES: Slippersnails are arched with a conspicuous ventral shelf.

Spotted slippersnails *(Crepidula maculosa)* have a shelf with a straight edge angling away from the apex. Most have brown spots on white.

Atlantic slippersnails *(C. fornicata)* have a coiled apex bent to one side, a smooth exterior, and a shelf with an indented edge. Color varies widely.

HABITAT: Spotted slippersnails are most likely to be found on offshore reefs. Atlantic slippersnails live in shallow waters on rocks and on other shells.

DID YOU KNOW? Slippersnails grow where they settle as tiny "spat" and have shell shapes that conform to their location. Atlantic slippersnails are famous for growing in stacks. The bottom snail in a stack began life as a male and switched to female. The snail arriving to grow on the bottom female remained male until another snail settled on it. Each arriving young snail assumes a male's role until another snail arrives, a process that can continue to the height of ten or more slippersnails. The stacks do function in reproduction, but the snail's species name may innocently refer to its curved shape. *Fornix* is Latin for arch.

Slippersnails
(Eastern White, Convex, and Spiny)

White Slippersnail

Convex and Spiny
Slippersnails

RELATIVES: Other slippersnails and cup-and-saucer snails. Each is within the family Calyptraeidae.

IDENTIFYING FEATURES:

Eastern white slippersnails *(Crepidula plana)* are white, thin, and flattened with a small, pointed apex.

Convex slippersnails *(C. convexa)* are small and deep shelled with an apex that curls. They are typically brown or purplish-brown with occasional dark spots or streaks. The deeply set shelf is white.

Spiny slippersnails *(C. aculeata)* differ in having roughened, sometimes spiny, radiating ridges. They are brownish with fine white rays.

HABITAT: These species live in shallow water where shell rubble is common. Eastern white slippersnails prefer to live inside vacated shells, especially pen shells and whelks.

DID YOU KNOW? All our slippersnails begin life as males and strategically switch sex based on environmental conditions. They make their living by filtering tiny food bits from the water. Slippersnails stay attached using the suction from their fleshy foot. One of their only travel opportunities is to have the old shell they live on hauled around by a hermit crab roommate, which is a common occurrence.

Eastern white slippersnail, max 1 in (2.5 cm)

Convex slippersnail, max 0.5 in (1.3 cm)

Spiny slippersnail, max 1 in (2.5 cm)

Striate cup-and-saucer, max 1.5 in (3.5 cm)

Brown baby ear, max 2 in (5 cm)

White baby ear, max 2 in (5 cm)

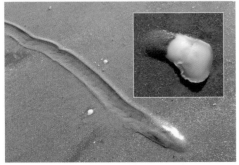

A living white baby ear tunnels through wet sand

Cup-and-Saucer Snail and **Baby Ears**

Cup-and-Saucer   Brown Baby Ear   White Baby Ear

RELATIVES: Cup-and-saucer snails are with slippersnails in the family Calyptraeidae. Baby ears share the family Naticidae with moonsnails.

IDENTIFYING FEATURES:

Striate cup-and-saucer snails *(Crucibulum striatum)* are shaped like a conical cap with a smaller cap attached within. Their apex is slightly turned and their exterior is lined with radiating ridges.

Brown baby ears *(Sinum maculatum)* have a large body whorl and gaping aperture but have a low spire like a flattened moonsnail. Their body whorl is sculptured with broad spiral grooves. Brown baby ears differ from white baby ears in having brown smudges and a higher, slightly pointed spire.

White baby ears *(Sinum perspectivum)* are similar to brown baby ears but have a more flattened spire. Shells are dull white or stained. Live animals have a white body enveloping their shell and look like a poached egg.

HABITAT: Brown baby ears live offshore. White baby ears are common within intertidal sands where their burrowing trails stand out at low tide.

DID YOU KNOW? Baby ear snails cannot withdraw their large foot into their shell. These species slide just beneath the surface of silty sands to prey on buried bivalves.

Moonsnails

Colorful and White Moonsnails   Northern Moonsnail   Miniature Moonsnail

RELATIVES: Moonsnails share the family Naticidae with baby ears and shark eyes.

IDENTIFYING FEATURES: All have a large, round body whorl, gaping aperture, and low, smooth spire.

Colorful moonsnails *(Naticarius canrena)* have a deep umbilicus half-filled with a traguslike pad (callus). They are creamy white with brown zigzags that are faded in old beach shells.

White moonsnails *(Polinices uberinus)* have no spiral cords over their glossy white exterior and have their umbilicus almost completely filled by a callus.

Northern moonsnails *(Euspira heros)* are large with a deep umbilicus. Their color is typically bluish-gray.

Miniature moonsnails *(Tectonatica pusilla)* are similar to white moonsnails but are smaller, rounder, and tan.

HABITAT: Northern moonsnails live in nearshore waters. The other moonsnail species live offshore.

DID YOU KNOW? These moonsnails prey on other mollusks by rasping drill holes into their victim's shells. They generally prowl along the bottom after dark and are able to "smell" the proteins given off by their prey. Living colorful moonsnails have a lovely patterned shell and an enormous brown-streaked foot spreading ten times their shell size.

Colorful moonsnail, max 2 in (5 cm)

White moonsnail, max 0.75 in (2 cm)

Northern moonsnail, max 4 in (10 cm)

Miniature moonsnail, max 0.25 in (6 mm)

13

Shark eye, max 3 in (7.5 cm)

The brown pad covering the umbilicus is the callus

Shark eye, size and color variation

Shark Eye

RELATIVES: Shark eyes (family Naticidae) are related to naticas, moonsnails, and baby ears.

IDENTIFYING FEATURES:

Shark eyes *(Neverita duplicata)*, or Atlantic moon snails, have a gaping aperture and a large body whorl that forms a smooth dome with a low spire. In many shells, an azure band on the lower whorls spirals inward to form a blue "eye." Color of the eye may also be purple, chestnut, or orange. The umbilicus is nearly covered by a brown, tragus-like pad (callus). Background shell color is tan, pinkish, brown-gray, blue-gray, or faded. The base of the shell is pale. The snail's thin operculum (aperture covering) is translucent amber. Shells in the northern end of our range tend to be browner with a more conical spire.

HABITAT: Offshore and in sandy shallows. Live shark eye snails are common in the swash zone during low tide off beaches with silty sands and protective shoals. The amber **opercula** from dead snails persist in the beach's wrack line.

DID YOU KNOW? Shark eyes plow through surf-zone sands in search of clams. Unlucky clams are enveloped by the snail's foot while an acidic secretion softens the clam's shell. A tooth-studded tongue (radula) then rasps a beveled hole (p. 70) at the softened spot. This hole allows a visit from the snail's proboscis, which injects enzymes to digest the clam's adductor muscles. With no muscles to hold it closed, the clam opens, allowing the shark eye to complete its meal of clam soup. The snail's favored diet includes surf clams (p. 52) and coquina clams (p. 55). Juvenile snails eat small clams, and larger adults eat large clams, each at a rate of almost a clam a day.

When plow-prowling for clams, shark eyes detect their prey like other moonsnails by "smelling" for telltale clam proteins. The clams are also able to detect the predatory snails and may flee to the sand's surface during a slow-motion attack. During a prowl, most of the snail's body is out of its shell and inflated with seawater. When picked up by a curious beachcomber, the snail must squirt out this water before it can withdraw into its shell and close its operculum.

A shark eye's amber operculum

Shark eyes breed in the surf zone by cementing their eggs with sand into a curled, gelatinous ribbon that cures into a rubbery **sand collar**. A circular opening atop the sand collar is where the snail's aperture was positioned as the collar formed. The collar is a study in hydrodynamics, being just the right shape to remain upright on shifting surf sands. To many, the shape suggests that the item was manufactured by humans and might be a discarded fragment of plastic. But a close examination reveals that the material of the collar is not uniformly molded and has thousands of transparent pockets. These pockets are the minute eggs, embedded within a single-layered matrix of sand grains cemented in gelatin. The collars disintegrate when eggs hatch, so whole collars found in the swash zone probably contain developing little snails.

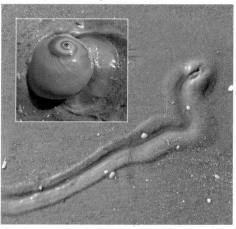

Shark eye tracks. Snail with exposed foot (inset)

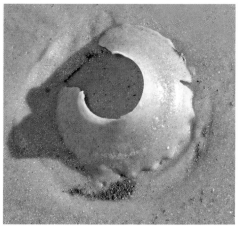

Shark eye "sand collar" eggs, max 4 in (10 cm)

Atlantic deer cowrie, max 5 in (13 cm)

Atlantic yellow cowrie, max 1.2 in (3 cm)

Coffeebean trivia, max 0.75 in (2 cm)

One-tooth simnia, max 0.75 in (2 cm)

Cowries, Trivia, and Simnia

Deer and Yellow Cowries and Trivia

One-tooth Simnia

RELATIVES: Cowries (family Cypraeidae) are distantly related to trivias (family Triviidae) and simnias (Ovulidae).

IDENTIFYING FEATURES:

Atlantic deer cowries *(Macrocypraea cervus)* have glossy, egg-shaped shells with a body-length, grinning aperture. Colors are chocolate with solid white spots or hazy brown with light bands.

Atlantic yellow cowries *(Erosaria acicularis)* are shaped like deer cowries but have a granular yellow pattern with marginal brown spots.

Coffeebean trivias *(Pusula pediculus)* are cowrie-shaped with riblets between a back groove and the aperture. They have three pairs of brown spots.

One-tooth simnias *(Simnialena uniplicata)* are spindle-shaped with an aperture stretching between each of its pointed ends. Colors vary from dark purple to yellow.

HABITAT: Cowries and trivias live on offshore reefs. Simnias live almost exclusively on sea whips. All these shells are uncommon finds.

DID YOU KNOW? Cowries feed on algae and colonial invertebrates, and trivias feed on tunicates and soft corals. A one-tooth simnia matches the color of the sea whip it lives and feeds on. The Atlantic deer cowrie is the largest of the world's 190 cowrie species.

Giant Tun, Figsnail, and Triton

Giant tun shell, max 10 in (25 cm)

RELATIVES: Tun shells (family Tonnidae) are distantly related to figsnails (family Ficadae). Tritons are in the family Ranellidae.

IDENTIFYING FEATURES:

Giant tuns *(Tonna galea)* are almost spherical in shape with a wide aperture, prominent spiral ridges, and a plain cream or brown color. Most beach finds are in pieces.

Atlantic figsnails *(Ficus papyratia)* have delicately tapered shells with a low spire and are sculptured with fine spiral ridges. Colors range from cream to tan, sometimes with faint brown dots.

Giant hairy tritons *(Cymatium parthnopeum)* are recognized by their thick, wavy outer lip, which is mahogany inside with white teeth. Less beachworn shells have a thick, brown periostracum (fuzz). The related Poulsen's triton *(C. cingulatum)* has a thin, white aperture and is less fuzzy.

HABITAT: Giant tuns, figsnails, and tritons live on offshore hard bottom.

DID YOU KNOW? Giant tuns feed on other mollusks, sea cucumbers, and fishes by engulfing their prey within a large expandable proboscis. Figsnails feed on sea urchins, and in life their shells are covered by a large, soft mantle. Tritons are rare finds on beaches; they live on offshore reefs where they are predators of varied invertebrates.

Atlantic figsnail, max 5 in (13 cm)

Giant hairy triton, max 3.5 in (9 cm)

17

Scotch bonnet, max 4 in (10 cm)

Clench helmet, max 12 in (30 cm)

Reticulate cowrie-helmet, max 3 in (7.6 cm)

Scotch Bonnet and Helmet Snails

RELATIVES: These gastropods are in the family Cassidae, distantly related to tun shells and figsnails.

IDENTIFYING FEATURES: All have a large body whorl with wide, toothy, grinning (or smirking) apertures.

Scotch bonnets *(Semicassis granulata)* have light shells with spiral grooves and a pointed spire. Colors range from white to cream with dark squares. The oldest beached shells are the most faded.

Clench (cameo) helmets *(Cassis madagascariensis)* have heavy shells with a low spire and a glossy, triangular aperture shield. Their whorl shoulders are studded with blunt knobs. Aperture teeth are light on chestnut.

Reticulate cowrie-helmets *(Cypraecassis testiculus)* have dense, egg-shaped shells with smooth spiral grooves and growth lines, and a rounded spire. They are chestnut to salmon with darker, blurry squares.

HABITAT: Offshore reefs. Shells are rarely beached except after storms.

DID YOU KNOW? North Carolina designated the scotch bonnet its state shell to honor its Scot forebears. The shell's color and shape almost resemble the plaid, woolen tam-o'-shanter cap worn by Scottish peasants. Scotch bonnets and helmet shells feed on sand dollars and sea urchins. Clench helmet populations are low and declining, perhaps due to trawling and habitat loss.

Nutmeg and Cantharus Snails

Common nutmeg, max 1.7 in (4.5 cm)

RELATIVES: Nutmegs (family Cancellariidae) and cantharus snails (family Buccinidae) are distantly related to auger snails.

IDENTIFYING FEATURES:

Common nutmegs *(Cancellaria reticulata)* have egg-shaped shells with a cross-hatched texture and whorls indented at the sutures. The inner lip of the aperture has two white folds on the columella. Shell colors vary between cream white and tan with blurry brown streaks.

Tinted canthari *(Gemophos tinctus)* have a similar shape to nutmegs but without distinct whorl indentations. The outer lip is toothed and the columella is glossy. Background shell color is cream or bluish-gray. Most have streaks and smudges of brown.

Ribbed canthari *(Hesperisternia multangula)*, also called false drills, have large ridges that are sharply angled at the whorl shoulders.

HABITAT: These snails live in sand, rubble, and seagrass to moderate depths.

DID YOU KNOW? Common nutmegs feed on soft-bodied animals buried in the sand. Canthari get their name from the cantharus, sacred cup of Bacchus, Roman god of wine. Cantharus snails prey on worms, barnacles, and other attached invertebrates.

Tinted cantharus, max 1.2 in (3 cm)

Ribbed cantharus, max 1.2 in (3 cm)

19

Eastern auger, max 2.4 in (6 cm)

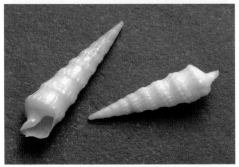

Concave auger, max 1 in (2.5 cm)

Sallé's auger, max 1.5 in (3.8 cm)

Sharp nassa, max 0.5 in (1.3 cm)

Augers and Sharp Nassa

Eastern Auger *Concave Auger and Sharp Nassa* *Sallé's Auger*

RELATIVES: Augers are allied in the family Terebridae. Sharp nassas share the family Nassariidae with mudsnails.

IDENTIFYING FEATURES: Augers have glossy, cone-shaped shells with short, distinct siphon canals.

Eastern augers *(Terebra dislocata)* are gray or orange-white with beaded spiral bands between whorls.

Concave augers *(T. concava)* resemble eastern augers except for the concave auger's namesake smooth, indented whorls. Color is white to yellow-gray.

Sallé's augers *(Hastula cinera)* are purple-gray with darkly streaked ribs below each whorl suture.

Sharp nassas *(Nassarius acutus)* have glossy, oval shells with conical spires. Their whorls have pointed beads connected by brown spiral lines.

HABITAT: Sandy shallows.

DID YOU KNOW? Augers feed on worms. Sallé's auger is an active hunter with a long stride and quick pace, nearly one "footstep" per second. It lunges when it finds a worm above the sand. Like other augers, Sallé's subdues its prey with a stab from a venomous, radular tooth. Their summer mating swarms are in the style portrayed by Burt Lancaster and Deborah Kerr in *From Here to Eternity,* with embracing pairs rolling in the swash zone.

Mudsnails

RELATIVES: Mudsnails and nassas are in the family Nassariidae.

IDENTIFYING FEATURES: These are small oval snails with conical spires.

Bruised nassas *(Nassarius vibex)* are light gray to dark with strong rounded axial ribs and a pointed spire. Their inner aperture lip is thickened by a wide glossy callus, which in darker shells bears a purple bruise.

Threeline mudsnails *(Nassarius trivittatus)* have yellowish-gray shells with shouldered (stepped) whorls and a woven, basketlike texture.

Eastern mudsnails *(Nassarius obsoletus = Ilyanassa obsoleta)* have solid brown shells with smooth, slanting, axial ribs. Their apex is typically worn. Live snails swarming over tidal flats are darkened by mud and algae.

HABITAT: Bruised nassas live in shallow seagrass and threeline mudsnails live on sandy bottom to moderate depths. Eastern mudsnails live over silty intertidal sands.

DID YOU KNOW? In Latin, *nassa* means wicker basket. Nassas and mudsnails eat algae, invertebrate eggs, carrion, and other easily outrun items. Eastern mudsnails swarm by the thousands where there is abundant food. They deposit their eggs in bristly capsules attached to hard surfaces, including other mollusk eggcases. Each 1-mm capsule contains several mudsnail eggs.

Bruised nassa, max 0.75 in (2 cm)

Threeline mudsnail, max 0.9 in (2.2 cm)

Eastern mudsnail, max 1.2 in (3 cm)

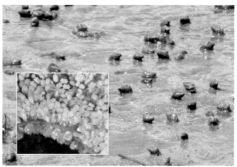

Eastern mudsnail swarm. Eggs, each 1 mm (inset)

21

Banded tulip, max 4 in (10 cm)

True tulip, max 5 in (13 cm)

Tulip snail egg capsules. Snails within (inset)

Tulip Snails

RELATIVES: Tulip snails (family Fasciolariidae) are related to spindle shells.

IDENTIFYING FEATURES: Tulip shells are shaped like pointed spindles with rounded curves and a stemlike siphon canal.

Banded tulips *(Fasciolaria lilium hunteria)* have cream to light gray shells with orange or gray splotches and fine spiral lines of reddish-brown. Their whorls are smooth.

True tulips *(Fasciolaria tulipa)* resemble banded tulips but have darker brown (or orange) splotches and interrupted, closer-set, spiral lines. Their whorls also differ in having fine ridges below each suture.

HABITAT: Banded and true tulips live on sand in water less than 100 ft (30 m).

DID YOU KNOW? True tulips prey on banded tulips, as well as on pear whelks and other mid-size gastropods. Tulip snails crawl into shallow waters during the winter to attach their clustered **egg capsules** to rocks, wood, and other bottom substrates. The egg capsules look like bouquets of smooth, flattened cones with frilly outer edges. Young tulip snails look like adults in miniature, and emerge from holes at the frilly end of each capsule. The capsules are formed of a tough, fingernail-like protein. If they rattle, the capsules are likely to contain tiny tulip shells. Several occupy each capsule.

Florida Horse Conch

Adult Florida horse conch shell, max 19 in (48 cm)

RELATIVES: Horse conchs are spindle shells, which share the family Fasciolariidae with tulip snails.

IDENTIFYING FEATURES:

Florida horse conchs *(Pleuroploca gigantea)* are thick-shelled with knobbed whorls that form a pointed spire about half the total shell-length. Adults are unmistakably large with a whitish spire and are often covered with brown, flaky periostracum. Beach-worn adult shells are white with a glossy tan interior. Living horse conchs have an orange-red body and a thick operculum (aperture cover). A **juvenile horse conch** resembles the adult but is a lighter, more uniform, peach-gold. Dark fossil shells are common on some beaches.

HABITAT: Horse conchs prefer silty sand in waters as shallow as the low-tide line out to moderate depths. They may be most common near inshore oyster beds.

Horse conch juveniles

DID YOU KNOW? As the largest snail in North America, horse conchs are able to prey on big gastropods, including large whelks and other horse conchs. Their egg masses comprise dozens of flattened bugles clustered in a twisted clump. Unlike tulip snail capsules, each buglelike cone has multiple lateral ridges. Like tulip snail capsules, the young escape by dissolving plugs to open holes at the end of each capsule.

Horse conch egg capsules. Snails within (inset)

Knobbed whelk, max 9 in (23 cm)

Kiener's whelk, a knobbed whelk variant

Knobbed whelks vary widely

Knobbed Whelk

RELATIVES: These snails share the family Melongenidae with other whelks.

IDENTIFYING FEATURES:

Knobbed whelks *(Busycon carica)* have large body whorls with distinct shoulders and a wide aperture tapering into their long siphon canal. Their shells are heavy and have several triangular knobs on the shoulder of the body whorl. Juveniles have brown stripes that fade as they become old adults, which often develop a glossy orange aperture. Background shell color is light gray to gray-brown. A variant form of knobbed whelk is the **Kiener's whelk**, which has a swelling at the lower body whorl and has shoulder spines that are extra large and recurved. Beach-worn knobbed whelk shells often remain intact although every edge is smoothed by surf-sanding.

HABITAT: Estuaries and nearshore waters out to depths of 150 ft (46 m). In estuaries, knobbed whelks can be common in oyster beds.

DID YOU KNOW? The knobbed whelk is the state shell of both Georgia and New Jersey, two states that are boundaries of the range where the species is most common.

These big snails feed on large bivalves like clams, oysters, and arks. The whelks use their sharp aperture lip to pry open the shell of their prey. During the sluggish attack, the whelk chips away at the tightly closed valves until an opening allows it to wedge its shell between the victim's valves and insert its foot. With a foot in the door, the whelk can eat. Many whelk shells show chipped apertures from this style of feeding.

Living knobbed whelks are most common in the intertidal zone during their two breeding seasons, which peak in late September and late April. Whelks larger than fist-size are mostly females, which are often pursued by one or more smaller males. Females grow faster than males, and reach maturity in about six years.

Like other whelks, knobbed whelks produce egg masses composed of dozens of disk-shaped capsules connected along a string. Disks at the earliest end of the string are small, contain no eggs, and serve as an anchor, often to a bit of shell rubble. The remainder of the disks each contain a dozen or more eggs. Egg-mass disk shape identifies the whelk species responsible. Knobbed whelk egg disks have simple, angled edges. Tiny hatchling whelks emerge from their egg capsules in about six weeks (Georgia) to six months (northern North Carolina). Egg-mass strings on beaches typically have holes indicating that the young snails have escaped, but some capsules rattle with the shells of those left behind.

A trawl fishery for knobbed whelks began in the nearshore waters of Georgia and South Carolina in the early 1980s. The meat was sold commercially for use in "conch" salads, chowders, fritters, and scungilli dishes. A peak in harvest occurred in the late 1980s, after-which the fishery collapsed. These states now have catch limits and hope to prevent overharvesting.

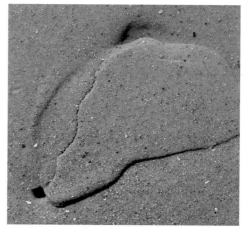

Whelk outline at low tide. Siphon hole (lower left)

A male knobbed whelk at low tide

Knobbed whelk egg mass string. Occupants (inset)

Lightning whelk, max 16 in (40 cm)

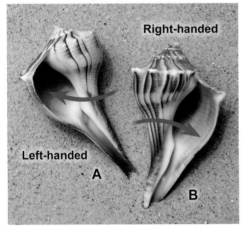

Right-handed

Left-handed

A

B

Lightning whelk (A), knobbed whelk (B)

Egg masses, with closeup of disks (inset)

Lightning Whelk

RELATIVES: These snails share the family Melongenidae with other whelks.

IDENTIFYING FEATURES:

Lightning whelks *(Busycon sinistrum)* have large body whorls with distinct shoulders bearing a dozen or more petite knobs. Like other whelks, they have a wide aperture tapering into their long siphon canal. The background color is cream or gray, with shells from younger whelks showing purple-brown axial streaks that look like lightning-bolts. Their shell is similar to the **knobbed whelk** (p. 24) except for being conspicuously left-handed, which puts the lightning whelk's body aperture (and body) on the left as it travels forward (spire in the rear). Most living lighting whelks have a charcoal-colored body, darker than knobbed whelks, which are cream.

HABITAT: Estuaries and nearshore sandy shallows.

DID YOU KNOW? The species name of the lighting whelk, *sinistrum,* describes the snail's left-handedness in Latin, not any malevolence. Nearly all other marine snails are right-handed. The same genes dictating left or right coiling in snails also govern our own asymmetry, such as the way our intestines coil. The **egg masses** strung together by lightning whelks have unique membranous edge projections that resemble plastic molding flash.

Channeled Whelk

RELATIVES: These snails share the family Melongenidae with other whelks.

IDENTIFYING FEATURES:

Channeled whelks *(Busycotypus canaliculatus)* have large body whorls and long apertures typical of whelks. Their angled whorl shoulders are edged by a lumpy spiral ridge, and a deep channel runs along the body-whorl suture well into the spire. Colors are gray to tan with only faint markings. The shell of a live channeled whelk is covered by a fuzzy, yellow-brown **periostracum,** which wears off in beached shells.

HABITAT: Channeled whelks live on intertidal sands and nearby shallows.

DID YOU KNOW? These snails scavenge and are often captured in baited traps. Whelks caught this way contribute to the scungilli served in Italian restaurants. The glossy white columella of this whelk was cut by Native Americans into beads that were strung into wampum belts, which were occasionally converted into colonial English currency. Whole shells were also used as drinking vessels. The whelk's sharp aperture edge, useful to the snail in prying open bivalve prey, provided a cutting tool for the earliest Americans. Channeled whelk egg masses are typical whelk "necklaces," but with individual disks having a unique, narrow, sharp-edged margin.

Channeled whelk, max 7.5 in (19 cm)

Live channeled whelk showing fuzzy periostracum

Channeled whelk egg disks. Young occupants (inset)

27

Pear Whelk

Pear whelk, max 5.5 in (14 cm)

Pear whelk egg disks. Young occupants (inset)

Pear (A) and channeled whelks (B) of similar size

RELATIVES: These snails share the family Melongenidae with other whelks.

IDENTIFYING FEATURES:

Pear (fig) whelks *(Busycotypus spiratus)* are similar in shape to the channeled whelk, but do not grow as large. The channel along the pear whelk's whorl suture is cut as with a "V" and disappears in earlier whorls near the spire's tip. The channel's cross section is U-shaped in the channeled whelk (p. 27). The pear whelk's angled shoulders are typically smooth. Colors are typically cream with brown, wavy, axial streaks. Live snails have a thin but fuzzy periostracum covering that wears off quickly after death.

HABITAT: Most common on muddy sand in shallow, quiet, bay waters.

DID YOU KNOW? Pear whelks feed on a variety of bivalves and would seem to compete for this food source directly with other whelks. But the large foot and rapid crawling speed of the pear whelk sets it apart in allowing capture of bivalves that flee (some bivalves just clam up to avoid predation). Pear whelks themselves commonly fall prey to stone crabs. This whelk's golden egg masses are smaller than those of the other local whelks. Its individual egg disks have narrow projections that stem from weakly angled corners. As in other whelk egg chains, the tiny disks at one end contain no eggs and are used to anchor the egg mass in the sand.

Dovesnails

Greedy, Well-ribbed, Lunar, and Fat Dovesnails   *West Indian Dovesnail*

RELATIVES: Dovesnails are in the family Columbellidae.

IDENTIFYING FEATURES: Dovesnails are small with short siphon canals and toothed aperture lips.

Greedy dovesnails *(Costoanachis avara)* have 12 smooth ribs on their body whorl, each with a white splotch.

Well-ribbed dovesnails *(Costoanachis lafresnayi)* resemble greedy dove snails but have straight-sided whorls that telescope into the spire and have spiral ridges between prominent ribs.

Lunar dovesnails *(Astyris lunata)* are plump, smooth, and brown with dark wavy lines or pale spiral bands.

West Indian dovesnails *(Columbella mercatoria)* have thick shells with revolving grooves. The narrow aperture has a thick outer lip with many teeth.

Fat dovesnails *(Parvanachis obesa)* have plump shells with distinct vertical ribs and fine revolving lines. The outer lip has a few small teeth inside.

HABITAT: Lunar dovesnails live intertidally. Remaining species are found in seagrass, rubble, or bryozoan colonies out to moderate depths.

DID YOU KNOW? These dovesnails are carnivores or scavengers. Dovesnail eggs are laid within single capsules, and young emerge either swimming or crawling.

Greedy dovesnail, max 0.75 in (2 cm)

Well-ribbed dovesnail, max 0.5 in (1.3 cm)

Lunar dovesnail, max 0.2 in (5 mm)

West Indian dovesnail, max 0.6 in (1.5 cm)

Fat dovesnail, max 0.25 in (0.6 cm)

29

Giant eastern murex, max 7 in (18 cm)

Apple murex, max 4.5 in (12 cm)

A communal apple murex egg mass

Murices *(Giant Eastern and Apple)*

RELATIVES: Murices share the family Muricidae with drills and rocksnails.

IDENTIFYING FEATURES: Murex shells are highly sculptured with round apertures and tubular siphon canals.

Giant eastern murices *(Hexaplex fulvescens)* are turnip-shaped with a body whorl sculptured by about 8 axial ridges (varices), each bearing pronounced hollow spikes. Beached shells are white to gray and may have only worn knobs instead of spikes.

Apple murices *(Chicoreus pomum)* are cream with brown bands and have 3 lumpy varices per whorl. Their inner aperture lip has a thin, flared margin and a dark blotch opposite the siphon canal.

HABITAT: Giant eastern murices live on sand to about 325 ft (100 m). Apple murices inhabit nearshore waters as shallow as the intertidal zone.

DID YOU KNOW? These murices prey on bivalves by rasping holes in their shells. They prefer oysters. Both of these murex species take part in group spawning events where multiple females add their egg capsules to a communal mass. Females then remain nearby, do not feed, and presumably stand guard over the eggs. These egg masses can be dozens of times larger than an individual murex snail.

Pitted Murex
and **Florida Rocksnail**

Pitted Murex   *Florida Rocksnail*

RELATIVES: Murices and rocksnails are in the murex family, Muricidae.

IDENTIFYING FEATURES:

Pitted murices *(Favartia cellulosa)* have dull white shells with 5–7 lumpy varices per whorl and a narrow, upturned siphon canal.

Florida rocksnails *(Stramonita hae-mastoma)* have sculptured shells and wide apertures with a toothed outer lip. They are whitish to grayish and frequently show red-brown spots. Their shells are highly variable, but all have spiral cords and axial ridges that are most prominent at the shoulders, which may have knobs. The outer lip interior may have brown lines or fine white ribs.

HABITAT: Pitted murices live near oyster and mussel beds. Rocksnails are most common near jetties and nearshore hard bottom.

DID YOU KNOW? Rocksnails feed on bivalves, gastropods, and barnacles. Their eggs are contained in tan or purple-stained, vase-shaped capsules that are attached to rocks during communal gatherings of snails.

Pitted murex, max 1 in (2.5 cm)

Florida rocksnail, max 3 in (8 cm)

Florida rocksnail,

31

Atlantic oyster drill, max 1.5 in (3.8 cm)

Thick-lip drill, max 1.3 in (4 cm)

Thick-lip drill egg capsules on an ark shell

Drills

RELATIVES: Drills are in the murex family, Muricidae.

IDENTIFYING FEATURES:

Atlantic oyster drills *(Urosalpinx cinerea)* have rounded shoulders and 9–12 rugged ribs per whorl. The aperture is oval with an open siphon canal. Colors range between yellow, orange, gray, and white, occasionally with brown streaks.

Thick-lip drills *(Eupleura caudata)* are pinkish with a long, thin siphon canal, an oval aperture, and a thick, toothed, outer lip opposite an equally thick ridge (varix) on the body whorl.

HABITAT: These drills live in estuaries within rocky, intertidal areas and oyster bars. Because of unintentional transport in ships' ballast water and in oyster seed beds, these drills now occur far outside their natural range, from England to the US Pacific.

DID YOU KNOW? Drills pierce young oysters by secreting acids onto their prey's shell and rasping the softened spot with a toothy radula. The resulting hole tapers to a small pinpoint, just wide enough for the drill to insert digestive enzymes and withdraw oyster soup. Female drills will attach dozens of flattened, leathery, urn-shaped **egg capsules** to a hard surface. Capsules are transparent and about half the length of the snail. Thick-lip drills lay capsules with a tiny side projection, and Atlantic oyster drill egg capsules are smooth.

Olive Shells and Marginella

Lettered Olive

Dwarf Olive and Marginella

Lettered olive, max 2.7 in (7 cm)

RELATIVES: Olive shells are in the family Olividae, distantly related to marginellas, family Marginellidae.

IDENTIFYING FEATURES: All have glossy shells with elongate apertures.

Lettered olives *(Oliva sayana)* have a thick shell with a small pointed spire. Newer shells are covered with slightly blurry, brown zigzags.

Variable dwarf olives *(Olivella mutica)* are gray to brown, variably marked, and have a spire extending half their shell length. The aperture is triangular with an inner, ridged, parietal callus that extends beyond the aperture up to the next suture. Live snails leave conspicuous tracks on sandy, low-tide flats. It is commonest of several *Olivella* species.

Common Atlantic marginellas *(Prunum apicinum)* have a glossy, egg-shaped shell with a low spire and a thick, smooth, outer aperture lip margin extending up past the preceding whorl. Colors range from gray to tan.

HABITAT: Silty sand in waters as shallow as the low-tide mark.

DID YOU KNOW? These snails cover their shells with their sensitive mantle, which requires comfortably smooth, porcelain shell finish. Each species will scavenge, but lettered olives prey mostly on coquina clams in the surf zone. In 1984, South Carolina designated the lettered olive as its state shell.

Variable dwarf olive, max 0.6 in (1.6 cm)

A live variable dwarf olive leaves a snaking trail

Common Atlantic marginella, max 0.5 in (1.3 cm)

33

Channeled barrel-bubble, max 0.25 in (6 mm)

Straight needle-pteropod, max 0.4 in (1 cm)

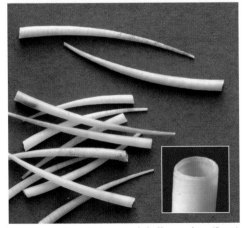

Straight ivory tuskshell, max 2 in (5 cm)

Bubble Shell, Pteropod, and **Tuskshell**

RELATIVES: Bubble shells (family Bullidae) are opisthobranchs (a group of gastropods containing sea slugs). Pteropods are in the order Thecosomata. Tuskshells are scaphopods (a separate class from the gastropods and bivalves) in the family Dentaliidae.

IDENTIFYING FEATURES:

Channeled barrel-bubbles *(Acteocina canaliculata)* have a fragile shell with an aperture along two-thirds of the shell length and the earliest whorls submerged into the spire. The similar Candé's barrel-bubble *(Acteocina candei)* has straight-sided whorls into a conelike spire.

Straight needle-pteropods *(Creseis acicula)* have glassy, needle-shaped shells that occasionally beach in massive numbers. This tiny, shelled sea slug has paired winglike flaps for swimming.

Ivory tuskshells (*Graptacme eborea*) are curved, tapered tubes, open at each end. The foot and mouth of the living tuskshell were formerly located at the wide end, which is smooth and round.

HABITAT: Bubble shells live in bays and sounds. Pteropods inhabit the surface of the open ocean. Ivory tuskshells live in sandy bottom offshore.

DID YOU KNOW? Pteropods feed by trapping plankton in a mucous web. Tuskshells live with their wide (anterior) end in the seabottom where they use their oral tentacles to feed on tiny forams.

Paper Nautilus
and **Ram's Horn Squid**

RELATIVES: These are cephalopods, a separate mollusk class from the gastropods, bivalves, and scaphopods. Argonauts are a type of octopus in the family Argonautidae. The ram's horn squid is in family Spirulidae.

IDENTIFYING FEATURES:

Paper nautili are the eggcases of the **greater argonaut** *(Argonauta argo)*, a species of open-ocean octopus. The female argonaut creates this paper-thin eggcase, which coils around the octopus similar to a wrinkled nautilus shell.

Ram's horn squid shells *(Spirula spirula)* are beached as white, chambered coils. In life, the coil is within the back end of the squid opposite its two large eyes and ten tentacles. The coil takes up almost half the squid, minus outstretched tentacles.

HABITAT: Argonauts and ram's horn squid live in the deep open ocean.

DID YOU KNOW? Paper nautili are not true shells because they are not produced by the mantle. The eggcase hardens from carbonates and proteins secreted by the webbing of two of the female's arms and functions only in egg protection. The argonaut holds herself within the case, but is not attached to it. Ram's horn squid use their buoyant, chambered coil to suspend themselves head-down in the water column. For protection, the squid puckers up by withdrawing its head and tentacles.

Paper nautilus, max 12 in (30 cm)

Living female argonaut stranded with her eggcase

Ram's horn squid shell, max 1 in (2.5 cm)

35

Blood ark, max 3 in (7.5 cm)

Transverse ark, max 1.4 in (3.6 cm)

Ponderous ark, max 2.8 in (7 cm)

White miniature ark, max 0.9 in. (2.2 cm)

Arks *(Blood, Transverse, Ponderous, and White Miniature)*

Blood, Transverse, and Ponderous Arks

White Miniature Ark

RELATIVES: Ark shells are allied within the family Arcidae.

IDENTIFYING FEATURES: Arks have thick shells with forward umbones and distinct ribs. In life, arks are coated with a brown, fuzzy periostracum.

Blood arks *(Lunarca ovalis)* have an oval shell with an arched hinge-line bearing about 7 teeth in front of the umbo and about 30 behind. Rear hinge teeth are largest and angled backward.

Transverse arks *(Anadara transversa)* have an elongate oval shell with a relatively straight hinge-line bearing mostly vertical teeth below a thin ligament scar the length of the hinge.

Ponderous arks *(Noetia ponderosa)* have a very thick, triangular shell with flat, divided ribs and an arched hinge-line below a broad moustachelike, grooved ligament area.

White miniature arks *(Acar domingensis)* have rear-pointed shells with heavy growth lines that give it a scaly look.

HABITAT: White miniature arks grow attached to offshore rubble. The other arks live in nearshore sands.

DID YOU KNOW? Robust shells make arks among the most common whole shells on high-energy beaches, where waves pulverize other mollusks. Blood arks have hemoglobin-red blood.

Arks *(Incongruous, Cut-ribbed, Mossy, and Turkey Wing)*

Incongruous ark, max 3 in (7.5 cm)

IDENTIFYING FEATURES:

Incongruous arks *(Scapharca brasiliana)* are thin-shelled for an ark. Their strong radial ribs have distinct beads like embossed dashes.

Cut-ribbed arks *(Anadara floridana)* have an elongate shell with a relatively straight hinge-line. The ribs are flattened, and each of the forward ribs (toward umbo) has a deep central "cut."

Mossy arks *(Arca imbricata)* have a straight hinge line almost the length of the shell. The ribs are beaded and the color is mostly chestnut brown.

Cut-ribbed ark, max 4.5 in (11.5 cm)

Turkey wings (zebra arks) *(Arca zebra)* are similar to mossy arks but have less-beaded ribs and a longer hinge line. Turkey wings have nested, red-brown Vs or Ws at the umbo that turn into oblique lines or zigzags rearward.

HABITAT: Incongruous arks live within sandy shallows, and cut-ribbed arks live in offshore sands. Mossy and turkey wing arks grow attached to near-shore rubble.

Mossy ark, max 2.5 in (6.3 cm)

DID YOU KNOW? The arched shell-gap opposite the umbo in mossy arks and turkey wings marks the opening where byssal threads anchor them to the bottom. The sand-dwelling arks use this attachment method only as juveniles. Arks feed like most bivalves, by inhaling water through a siphon and filtering out plankton with their gills.

Turkey wing, max 3.6 in (9 cm)

37

Giant bittersweet clam, max 4 in (10 cm)

Comb bittersweet clam, max 1.2 in (3 cm)

Asian green mussel, max 3.5 in (9 cm)

Bittersweet Clams
and Asian Green Mussel

Bittersweet Clams   Green Mussel

RELATIVES: Bittersweets (family Glycymerididae) may be distantly related to the arks. Mussels (family Mytilidae) are distantly related to penshells.

IDENTIFYING FEATURES: Bittersweet clams have heavy, rounded, circular shells with thick, arching hinge-lines bearing several prominent teeth on either side of the umbo.

Giant bittersweets *(Glycymeris americana)* have a small, rounded umbo and roughly 50 flattened ribs. They are glossy cream with concentric, blurry necklaces of tan or rust.

Comb bittersweets *(Tucetona pectinata)* have a relatively pointed umbo, and 20–30 raised ribs. They are slightly roughened by growth lines and are grayish-white with brown or purple spatters.

Asian green mussels *(Perna viridis)* have thin shells with a smooth, green and brown exterior. See other mussel features (p. 39).

HABITAT: These bittersweets live in sand out to moderate depths. Green mussels attach to shallow rubble and pilings.

DID YOU KNOW? Bittersweets taste as their name suggests. They live unattached and have light-sensitive eyespots along their mantle. Following accidental transport in ship-ballast water, green mussels have invaded our region from Asia and are pushing out native species.

Mussels *(Scorched and Ribbed)*

RELATIVES: Mussels (family Mytilidae) are distantly related to penshells.

IDENTIFYING FEATURES: Mussels have thin shells that fan out from their umbones and tend to retain their thin, shiny periostracum. Shell interiors are typically nacreous (lustrous, like pearls).

Scorched mussels *(Brachidontes exustus)* have shells with radiating ribs and 2–3 hinge teeth under the umbo.

Ribbed mussels *(Geukensia demissa)* have shells with radiating ribs and no hinge teeth. Shells without the brown periostracum are yellowed gray with occasional purple tinges.

HABITAT: These mussels grow attached to rocks, pilings, and rubble in waters as shallow as the intertidal zone. Live scorched mussels are abundant on rock jetties. Ribbed mussels are most common in saltmarsh.

DID YOU KNOW? Mussels form "beds" that are habitat for other animals. Scorched mussels are found in densities of nearly 5000 per square yard. Uncommonly for a bivalve, ribbed mussels spend considerable time above water at low tide and gape during this period in order to breathe. They have a close association with the roots of saltmarsh grasses and gain an important fraction of their food from cellulose sloughed from the plant.

Scorched mussel, max 1.5 in (4 cm)

Mass of scorched mussels growing on a jetty

Ribbed mussel, max 5 in (13 cm)

Ribbed mussels in old marsh exposed on a beach

39

Hooked mussel, max 2 in (5 cm)

Blue mussel, max 2.5 in (6.5 cm)

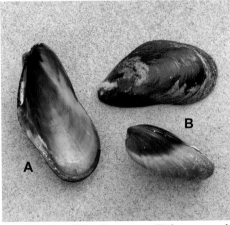

Southern (A) and American (B) horse mussels

Mussels *(Hooked, Blue, and Horse)*

Hooked and Horse
Mussels

Blue Mussel

IDENTIFYING FEATURES:

Hooked mussels *(Ischadium recurvum)* have a curved, triangular shell with radial ribs that branch toward the end opposite the umbo.

Blue mussels *(Mytilus edulis)* have no ribs but have fine concentric growth lines. Outer colors range from violet to black, and inner color is shiny blue.

Horse mussels *(Modiolus* spp.) have inflated shells and an umbo just shy of their upper end. **Southern horse mussels** *(M. squamosus)* reach 2.5 in (6.5 cm), have less-inflated umbones, and are whitish or purple after beach wear. The similar **American horse mussel** *(M. americanus)* reaches 4 in (10 cm), has bulbous umbones, and is bright red through its golden periostracum.

HABITAT: These mussels grow attached to rocks or pilings in estuarine waters.

DID YOU KNOW? These mussels attach to their substrate by byssal threads made of an elastic protein ten times the strength of our own tendons. A gland makes the threads by secreting a fluid that solidifies upon contact with water. As juveniles, mussels also secrete an enzyme with their foot that dissolves the mussel's thread attachments. In making new attachments, the mussel can move over rocks like a mountain climber. Hooked mussels are the principal food eaten by many marsh duck species.

Penshells

RELATIVES: Penshells (family Pinnidae) are distantly related to mussels.

IDENTIFYING FEATURES: Penshells have thin, amber-brown, fanlike valves.

Sawtooth penshells *(Atrina serrata)* have about 30 radiating ribs bearing hundreds of short, hollow prickles.

Sawtooth penshell, max 9 in (23 cm)

Half-naked penshells *(Atrina seminuda)* have about 15 radiating ribs bearing a few to dozens of long tubular spines. Their posterior (fan end) muscle scar is completely within their pearly (or cloudy) nacreous area.

Stiff penshells *(Atrina rigida)* are darker and broader than half-naked penshells and have their posterior **muscle scar** outside the shiny nacre.

HABITAT: Sawtooth and half-naked penshells live in colonies with individuals buried in soft sediment out to 20 ft (6 m). Stiff penshells live in bays and sounds.

Half-naked penshell, max 9 in (23 cm)

DID YOU KNOW? Penshells anchor themselves with golden byssal threads, which lead from their pointed (front) end to a small bit of rubble beneath the sand. Like most bivalves, they are filter feeders. Many living penshells have pale, soft-bodied shrimp or pea crabs living within their mantle cavity. Like pearl oysters, penshells occasionally produce pearls as a response to an irritant. These "pinna pearls" range from silvery to orange.

Stiff penshell, max 11 in (28 cm). Arrow shows scar

41

Lion's-paw, max 6 in (15.2 cm)

Rough scallop, max 1.5 in (4 cm)

Atlantic sea scallop, max 6.5 in (16.5 cm)

Scallops *(Lion's-paw, Rough, and Atlantic Sea)*

Lion's Paw and   Sea Scallop
Rough Scallop

RELATIVES: Scallops are allied within the family Pectinidae.

IDENTIFYING FEATURES: Scallops have round or oval shells with winglike "ear" projections from the umbo.

Lion's-paws *(Nodipecten nodosus)* have thick, flattened shells with 7–8 large, roughly ridged ribs bearing occasional hollow knuckles. Their outer shell color is commonly orange or brick red, but may range from pale to dark purple.

Rough scallops *(Lindapecten muscosus)* have about 19 ribs that are roughened by tiny spoon-shaped prickles. Beach-worn shells are less prickly. Most rough scallops are solid-colored lemon, peach, or tangerine, but some are mottled with plum.

Atlantic sea scallops *(Placopecten magellanicus)* are smooth with fine concentric lines. Upper valves are purplish, sometimes rayed, and lower valves are mostly white.

HABITAT: These scallops live offshore in sandy rubble.

DID YOU KNOW? Sea scallops are rare on the beach, but are the target of a fishery that dredges them from the sea bottom offshore. The scallop's adductor muscles (meats) are dissected out and the rest discarded at sea. The harvest has grown to over 30,000 tons of scallop meats per year in the US Atlantic.

Scallops *(Calico and Bay)*

Calico Scallop

Bay Scallop

IDENTIFYING FEATURES:

Atlantic calico scallops *(Argopecten gibbus)* have shells with 19–21 rounded ribs. Shell colors vary through white, yellow, orange, red, purple, and gray, generally with splotches of dark on light. Their ears are typically worn.

Atlantic bay scallops *(Argopecten irradians)* have 17–18 ribs that are squared in in comparison to calico scallops. Shell color is white, gray-brown, or orange.

HABITAT: Atlantic calico scallops live on sand bottom at depths to 1300 ft (400 m). Bay scallops live on muddy sands and seagrass in shallow waters.

DID YOU KNOW? Calico scallop fishing off North Carolina peaked in the 1980s, but diminished following a population crash from overharvest and habitat alteration. Pale scallop shells found on the beach are the right (lower) valves on which the scallop rests. To escape predators, a scallop can swim nine body-lengths per second by clapping its valves, which jets water out on either side of its hinge. In addition to people, scallops are eaten by gastropods, squid, octopodes, sea stars, and crabs. Bay scallops are also eaten by ring-billed and herring gulls, which pluck the scallops from the estuary at low tide, fly over a hard surface, and drop their catch to crack it open. This clever method carpets beachside parking lots with broken shells.

Atlantic calico scallop, max 2.7 in (7 cm)

Atlantic bay scallop, max 4 in (10 cm)

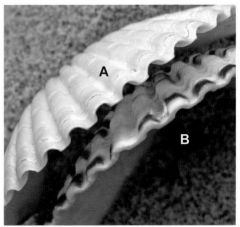
Bay scallop (A) and calico scallop (B)

43

Round-rib scallop, upper valve, max 2 in (5 cm)

Atlantic kittenpaw, max 1.2 in (3 cm)

Atlantic thorny oyster, max 5.1 in (13 cm)

Round-rib Scallop, Atlantic Kittenpaw, and Thorny Oyster

RELATIVES: Scallops are in the family Pectinidae. Kittenpaws (Plicatulidae) are distantly related to thorny oysters (family Spondylidae).

IDENTIFYING FEATURES:

Round-rib scallops *(Euvola raveneli)* have an upper valve that appears as if it were melted flat. This valve has round, separated ribs and varies from light gray to purple with rayed streaks. The lower valve is domed and white.

Atlantic kittenpaws *(Plicatula gibbosa)* have thick, tough, flattened shells with 6–10 curving, digitlike ribs. They are white to gray except for their tabby-orange ribs marked with numerous, thin, red-brown lines.

Atlantic thorny oysters *(Spondylus americanus)* have thick, circular, lumpy valves with occasional thorns (long in unworn shells). The hinge on the cup-shaped lower valve has two large cardinal teeth separated by a split, and the upper valve has two corresponding sockets. Colors are orange to brick red.

HABITAT: All live offshore in depths to 300 ft (91 m). Atlantic kittenpaws and thorny oysters live attached to rocks.

DID YOU KNOW? Left-valve kittenpaws are most common because the right valve often remains attached where the animal lived. The right valve retains an impression of the hard surface on which it grew.

Common Jingle
and **Crested Oyster**

RELATIVES: Common jingles (family Anomiidae) are distantly related to crested oysters (family Ostreidae).

IDENTIFYING FEATURES:

Common jingles *(Anomia simplex)* have round, brittle, pearly-translucent shells with no obvious hinge. Their colors include silver-gray, white, yellow, and orange. Black shells have been stained by sulfurous sediments. Right (lower) valves have a hole and no umbo (top left in upper image).

Common jingles, max 2 in (5 cm)

Crested oysters *(Ostreola equestris)* have lumpy, oval shells that are ruffled along the top edge in older specimens. Hinges have pimplelike teeth and the muscle scar almost central.

HABITAT: Common jingles and crested oysters live in shallow water attached to rocks, wood, and other shells.

DID YOU KNOW? Most beached jingle shells are the unattached left valve. In life, the mollusk remains attached to a hard surface by calcified byssal threads that stem from the hole in their lower valve. Although brittle, jingle shells are strong for how thin they are. Like other nacreous shells, strength comes from microscopically thin hexagonal platelets that are laid in offset layers like bricks. This keeps crack lines from spreading. When crushed, the thin layers break into bits that refract and reflect light like glitter.

Common jingles crushed into shell glitter

Crested oyster, max 2 in (5 cm). Hinge (inset)

45

Eastern oyster, max 6 in. (15 cm)

Eastern oyster shells are abundant on many beaches

Oyster bed from former marsh, exposed in surf

Eastern Oyster

RELATIVES: Eastern oysters are allied with crested and frond oysters in the family Ostreidae and are distantly related to scallops, jingles, limas, and pearl oysters.

IDENTIFYING FEATURES:

Eastern oysters *(Crassostrea virginica)* have lumpy shells that vary from oval to clown-shoe shapes. They have no hinge teeth, and their inner surface is smooth with a purple muscle scar. The lower valve is flat and often remains cemented to the surface where the oyster lived.

HABITAT: Estuarine waters, typically less saline than where crested oysters live. Eastern oysters attach to rocks, debris, or other oysters, in vast beds.

DID YOU KNOW? Eastern oysters were formerly super-abundant and a staple food of native Americans in the region. In the late 19th century, industrial oyster harvest from the central Atlantic coast took 27 million bushels each year. By 2004, harvest had declined 99%. In addition to a vast overharvest, oyster beds also suffered from introduced diseases and loss of attachment habitat due to dredging. Although their importance to the menu of seafood raw bars is obvious, oysters also play a critical role in creating habitat for a variety of other animals. Their filter-feeding cleans estuarine waters, and they are sensitive to poor water quality, making oysters excellent environmental sentinels.

Oysters *(Frond, Pearl, and Wing)* and **Antillean Fileclam**

Frond Oyster and Fileclam

Pearl-oyster and Wing-oyster

RELATIVES: Frond oysters (Ostreidae), pearl and wing oysters (Pteriidae), and fileclams (Limidae) are distantly related.

IDENTIFYING FEATURES:

Frond oysters *(Dendostrea frons)* have yellow- or purple-colored oval shells with strong radial ridges ending in interlocking scalloped margins. Those attached by clasping, fingerlike shell projections to the branches of soft corals have the most elongate shell shape.

Atlantic pearl-oysters *(Pinctada imbricata)* have valves with a straight hinge with short, triangular front and rear wings. Unworn shells have a scaly, fringelike periostracum.

Atlantic wing-oysters *(Pteria colymbus)* look similar to pearl oysters but have longer wings that extend past the rest of the shell.

Antillean fileclams *(Limaria pellucida)* have thin white shells with fine riblets.

HABITAT: Frond, pearl, and wing oysters live attached to many objects, including sea whips. Fileclams live in offshore crevices.

DID YOU KNOW? Both pearl and wing oysters occasionally produce tiny pearls. Fileclams live in a nest made of their own byssal threads and can swim away from predators by clapping their valves.

Frond oyster, max 1.5 in (4 cm)

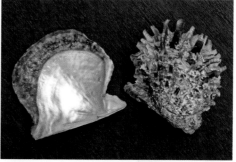

Atlantic pearl-oyster, max 3.5 in (9 cm)

Atlantic wing-oyster, max 3.5 in (9 cm)

Antillean fileclam, max 1.1 in (2.8 cm)

47

Buttercup lucine, max 2.5 in (6.4 cm)

Cross-hatched lucine, max 1 in (2.5 cm)

Pennsylvania lucine, max 2 in (5 cm)

Lucines

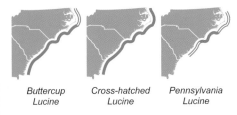

| Buttercup Lucine | Cross-hatched Lucine | Pennsylvania Lucine |

RELATIVES: Lucines are allied within the family Lucinidae.

IDENTIFYING FEATURES: Lucines have thick circular shells with forward-pointing umbones above a distinct, heart-shaped impression (the lunule) split by the valve opening.

Buttercup lucines *(Anodontia alba)* have a forward flare that forms a keel protruding more than the umbo. The outer shell is dull white with fine growth lines and the inner shell is butter-yellow or cream.

Cross-hatched lucines *(Divalinga quadrisulcata)* are moderately inflated with relatively thin valves sculptured by numerous, parallel lines that make the shell appear covered with fingerprints. Beached shells are glossy white or ivory.

Pennsylvania lucines *(Lucina pensylvanica)* have very thick, off-white valves with a deep furrow either side of the umbo. Thin, scaly, growth lines are separated by smooth bands.

HABITAT: All live in sandy shallows. Buttercup lucines and Pennsylvania lucines can live as deep as 300 ft. (90 m).

DID YOU KNOW? Lucines are named for Lucina, an aspect of the Roman goddess Juno who represented light and childbirth. These burrowing mollusks have a muscular foot that can extend six times their shell length.

Jewelboxes and Carditid

RELATIVES: Jewelboxes (family Chamidae) and carditids (family carditidae) are allied with clamlike bivalves.

IDENTIFYING FEATURES:

Florida spiny jewelboxes *(Arcinella cornuta)* are shaped like tubby commas bearing about 8 radiating ridges with hollow spines (or knobs, if beach-worn). They are white with a pinkish interior.

Leafy jewelboxes *(Chama macerophylla)* have thick, oval shells covered in numerous scaly ridges. Beach-worn shells are lumpy, but new shells may have long, hollow scales. They are generally yellow or chalky, but are often orange or lavender.

Corrugate jewelboxes *(Chama congregata)* have a corrugated exterior, fine ridges within the inner valve margins, and are reddish outside, purplish inside.

Three-tooth carditids *(Pleuromeris tridentata)* have small, thick triangular shells with beaded radial ribs. Colors are cream and rusty pink.

HABITAT: Jewelboxes live cemented to reefs and debris to moderate depths. Florida spiny jewelboxes detach when young to grow free within sandy rubble. Carditids live in shallow sands.

DID YOU KNOW? The spines and scales of jewelboxes are an important line of defense against being drilled by gastropod predators (p. 70).

Florida spiny jewelbox, max 2.5 in (6.3 cm)

Leafy jewelbox, max 3.1 in (8 cm)

Corrugate jewelbox, max 1 in (2.5 cm)

Three-tooth carditid, max 0.25 in (6 mm)

Atlantic strawberry-cockle, max 2 in (5.1 cm)

Atlantic giant cockle, max 5.2 in (13.2 cm)

Spiny papercockle, max 1.8 in (4.5 cm)

Florida pricklycockle, max 2.7 in (6.9 cm)

Cockles and **Florida Pricklycockle**

Strawberry-, Papercockle,
and Pricklycockle

Atlantic Giant Cockle

RELATIVES: Cockles are allied within the family Cardiidae.

IDENTIFYING FEATURES: Cockle shells are oval, inflated, and have a large umbo with one central tooth and socket.

Atlantic strawberry-cockles *(Americardia media)* are cream with red-brown specks and have flattened ribs that feel like sandpaper. An angled ridge runs across the longest part of the shell.

Atlantic giant cockles *(Dinocardium robustum)* are cream with brown or tan in segments along their ribs, which are rounded and bumpy along the shell's front, flattened and smooth in the rear.

Spiny papercockles *(Papyridea lata)* are compressed with rear ribs ending in protruding spines. They are mottled with pale pink, purple, orange, or red-brown. Similar *P. soleniformis* is more elongate and speckled brown on cream.

Florida pricklycockles *(Trachycardium egmontianum)* have about 30 ribs covered by strong scales (in unworn shells) and ending in a deeply serrated hind margin. External color is cream with tan or purple-brown splotches. Their valves inside are salmon and/or purple.

HABITAT: Sandy shallows off beaches.

DID YOU KNOW? Giant cockles are also called heart cockles because of the end-on profile that two valves make. Cockles are common in chowders.

Yellow Pricklycockle
and **Eggcockles**

Pricklycockle and Common and Painted Eggcockles

Yellow Eggcockle

Yellow pricklycockle, max 2.5 in (6.4 cm)

RELATIVES: Pricklycockles and eggcockles are in the family Cardiidae.

IDENTIFYING FEATURES:

Yellow pricklycockles *(Trachycardium muricatum)* have about 35 ribs with small scales. They have tinges of yellow inside and out and may tend toward peach with occasional red-brown streaks.

Common eggcockles *(Laevicardium laevigatum)* have valves with an oblique oval shape and ridges along the inner margin. They are glossy white or yellow with occasional rosy tinges. Older beached shells are white and less glossy.

Common eggcockle, max 3 in (7.6 cm)

Painted eggcockles *(Laevicardium pictum)* have a compressed, skewed triangular shape. They are cream with blurry, orange or brown zigzags.

Yellow eggcockles *(Laevicardium mortoni)* are almost evenly rounded with a central umbo and are colored by relatively distinct rows of brown, purple, or orange zigzags.

Painted eggcockle, max 1 in (2.5 cm)

HABITAT: Sandy shallows to offshore. Yellow eggcockles prefer shallow bays.

DID YOU KNOW? Prickles may help anchor pricklycockles in the sand. Eggcockles avoid predators differently—by leaping away using their muscular foot. This does not always allow yellow eggcockles to escape ducks, who have this bivalve on their favorite-foods list.

Yellow eggcockle, max 1 in (2.5 cm)

Southern surfclam, max 5.1 in (13 cm)

Atlantic surfclam 8 in (20.3 cm)

Surfclams
(Southern, Atlantic, and Dwarf)

Southern Surfclam | Atlantic Surfclam | Dwarf Surfclam

RELATIVES: Surfclams share the family Mactridae with duckclams.

IDENTIFYING FEATURES: These clams have a spoon-shaped pit behind the central hinge teeth.

Southern surfclams *(Spisula raveneli)* have a central umbo and fine growth lines. Colors are white to dirty cream with rusty tones. Dark shells are stained.

Atlantic surfclams *(Spisula solidissima)* look similar to southern surfclams, but have thicker, more inflated shells with a more bulbous umbo.

Dwarf surfclams *(Mulinia lateralis)* have an umbo forward of center and a tapered hind end. Colors may be white, cream, gray, or purple-gray, with highlighted growth bands.

HABITAT: All live in sand from just off the beach out to moderate depths (165 ft or 50 m). Dwarf surfclams are also common in shallow lagoons.

DID YOU KNOW? Atlantic surfclams are commercially harvested for food between Virginia and New Jersey. The abundant dwarf surfclam feeds a host of estuarine animals including ducks and a hefty shell-crunching fish called the black drum *(Pogonias cromis)*. As their name suggests, these bivalves are most common just outside the surf zone.

Dwarf surfclam, max 0.75 in (2 cm)

Fragile Surfclam, Duckclams, and **Atlantic Rangia Clam**

Fragile Surfclam and Duckclams   *Atlantic Rangia Clam*

Fragile surfclam, max 4 in (10.2 cm)

RELATIVES: These clams are allied with surfclams in the family Mactridae.

IDENTIFYING FEATURES: All have a spoon-shaped pit behind the central hinge teeth.

Fragile surfclams *(Mactrotoma fragilis)* are thin-shelled with a forward umbo. Shell color is cream with some remaining periostracum behind a ridgeline on the hind end.

Channeled duckclams *(Raeta plicatella)* have white, thin, ear-shaped shells with strong concentric growth ridges.

Channeled duckclam, max 3.2 in (8.1 cm)

Smooth duckclams *(Anatina anatina)* have thin, off-white, ear-shaped shells with smooth growth lines and a distinct ridge leading from the umbo.

Atlantic rangia clams *(Rangia cuneata)* have thick shells with an inflated umbo. Their front hinge tooth is large and rectangular and their rear tooth is a long, flat ridge. Beach shells are well worn.

Smooth duckclam, max 3 in (7.6 cm)

HABITAT: Fragile surfclams and channeled duckclams live in sand just outside the surf zone. Smooth duckclams live offshore. Atlantic rangia live in the muddy sands of brackish bays.

DID YOU KNOW? Each of these species lives buried with two united siphons that they extend just above the sand allowing the clam to filter-feed.

Atlantic rangia clam, max 2.7 in (7 cm)

Alternate tellin, max 2.7 in (6.9 cm)

Rainbow tellin, max .5 in (1.2 cm)

Texas tellin, max .5 in (1.2 cm)

Striate tellin, max 1 in (2.5 cm)

Tellins

RELATIVES: Tellins are allied with macomas in the family Tellinidae.

IDENTIFYING FEATURES: Like other tellins, each of these shells is less rounded in the rear where they bend right (outward in the right valve).

Alternate tellins *(Eurytellina alternata)* are shiny yellow-white with numerous concentric grooves between flattened concentric ridges. They may have yellow or pink radiating from the umbo.

Rainbow tellins *(Scissula iris)* have thin, translucent shells tinged with pink. The concentric growth lines are smooth.

Texas tellins *(Angulus texanus)* have opaque, white shells and are steeply sloped behind the hinge.

Striate tellins *(Merisca aequistriata)* have chalky shells with distinct growth lines and a strong radial ridge rearward.

HABITAT: Each of these tellins lives in sand off beaches out to moderate depths.

DID YOU KNOW? Tellins lie beneath the sand on their left valve so that their posterior curves upward. This accommodates their long intake siphon, which draws in surface morsels. They are deposit feeders rather than filter feeders. Their bladelike form and strong foot allow rapid burrowing should a predator approach. Like the arks, tellin bodies are red from the oxygen binding pigment, hemoglobin.

Coquina Clams

RELATIVES: Coquina clams (family Donacidae) are related to tellins.

IDENTIFYING FEATURES:

Variable coquina clams *(Donax variabilis)* have glossy, wedge-shaped shells with faint riblets. Grooved teeth line their inner margins. Patterns vary between solids, radial rays, and concentric bands, and may include any color.

Little coquina clams *(Donax fossor)* are similar to variable coquinas but have a rounder posterior end and umbo.

HABITAT: Variable coquinas live in the swash zone, and little coquinas are only slightly deeper.

DID YOU KNOW? Variable coquinas are one of the most abundant and ecologically important mollusks on our Southeastern beaches. Specialized for life in wave-washed sand, they filter-feed on algae and bacteria swept to shore and maximize their time in this zone by migrating back and forth with each tide. They are a critical food for shore birds and surf fishes, but would rather not be. Coquinas burrow rapidly to outrun predators. They also have a unique mutualistic relationship with a hydroid that grows attached to the clam's shell. Coquinas with hydroid colonies are less likely to be eaten by predatory moonsnails and fish. The hydroids also benefit; without the coquina, the colonies would have no attachment in the turbulent surf.

Living variable coquina clams, max 1 in (2.5 cm)

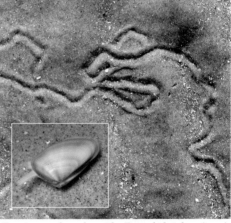

Coquina tracks at low tide. Extended siphon (inset)

Variable (A), little coquina (B, max 0.7 in, 1.7 cm)

Purplish semele, max 1.3 in (3.4 cm)

Cancellate semele, max 0.75 in (1.9 cm)

Atlantic semele, max 1.5 in (3.8 cm)

Tellin semele, max 1 in (2.5 cm)

Semeles

Purplish and Cancellate   Atlantic Semele   Tellin Semele

RELATIVES: Semeles (family Semelidae) are related to tellins and coquinas.

IDENTIFYING FEATURES: Like tellins, a semele's hind end bends right. Their hinges have a diagonal depression angling back from the umbo.

Purplish semeles *(Semele purpurascens)* have smooth oval shells with an umbo toward the rear. They have smudge-streaks of blurry purple, brown, or orange.

Cancellate semeles *(Semele bellastriata)* are cream or gray with concentric ridges and radial riblets front and rear.

Atlantic semeles *(Semele proficua)* have a central umbo and are cream with occasional nervous purple lines.

Tellin semeles *(Cumingia tellinoides)* are dirty white and have a distinct point at the rear shell. At their hinge, a spoon-like depression beneath the umbo protrudes into the inner shell.

HABITAT: Purplish and cancellate semeles live in sand banks off beaches out to moderate depths. Atlantic and tellin semeles prefer inlet areas and shallow bays open to the sea.

DID YOU KNOW? A semele's hinge depression (chondrophore) bears a cushiony pad that springs the valves open when the animal's adductor muscles relax.

Tagelus (Short Razor) Clams

Stout tagelus, max 3.9 in (10 cm)

RELATIVES: Tagelus clams (family Solecurtidae) are related to semeles and tellins.

IDENTIFYING FEATURES: These clams have central umbones and elongate shells that gape at each end.

Stout tagelus clams *(Tagelus plebeius)* have thick, lumpy shells with smooth growth lines. They are white, ivory, light gray, or purplish. Freshly beached shells have a greenish periostracum.

Purplish tagelus clams *(Tagelus divisus)* have smooth, thin shells that are tinted purple inside and out. A darker purple ray from the umbo marks a slightly raised internal rib. Small shells may have a covering of brown periostracum.

Stout tagelus burrows in former marsh mud

HABITAT: These clams live in the sand or mud of shallow estuaries. Stout tagelus clams prefer closed lagoons and purplish tagelus clams prefer bays open to the sea.

DID YOU KNOW? Tagelus clams feed on suspended particles and live in deep burrows with only their siphons exposed. Old stout tagelus burrows are often conspicuous in the peat and mud from former marsh exposed on eroding beaches. In the living marsh, tagelus clams often compose more than 90% of the biomass on mud flats.

Purplish tagelus, max 1.6 in (4.0 cm)

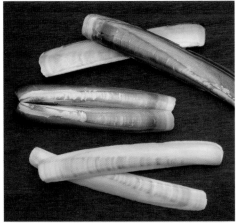

Minor jackknife clam, max 5 in (13 cm)

Mass stranding of minor jackknife clams

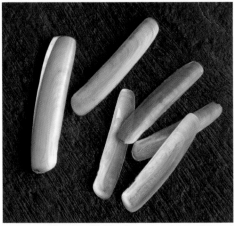

Green jackknife clam, max 1.5 in (3.8 cm)

Minor and Green Jackknife Clams

RELATIVES: Minor jackknife clams (family Cultellidae) are only distantly related to green jackknife clams (family Solenidae).

IDENTIFYING FEATURES:

Minor jackknife clams *(Ensis megistus)* have fragile shells with a curved straight-razor shape and are about nine times as long as they are wide. Shells are purplish inside and outside show a shell-length wedge of lavender growth bands against a whitish background.

Green jackknife clams *(Solen viridis)* have a straight upper-shell edge and are about four and a half times as long as they are wide.

HABITAT: Minor jackknife clams live burrowed into the sands of offshore shoals. Green jackknife clams live in the muddy sands of shallow sounds and bays.

DID YOU KNOW? Jackknife clams burrow vertically with astonishing speed and can dig to more than an arm-length's depth in the sand. To dig, they inflate their foot hydraulically, extend it down into the sand with the aid of squirting water, spread the foot into an anchor, and deflate the foot to pull the shell down. They can also swim. To do this, they fold their foot against the side of their shell and flick it backward. This action springs the clam forward. When rough seas erode offshore shoals, minor jackknife clams often strand on the beach in great numbers.

Venus Clams *(Quahogs)*

Southern Quahog

Northern Quahog

RELATIVES: Quahogs are venus clams in the family Veneridae, distantly related to false angelwing clams.

IDENTIFYING FEATURES: All venus clams have three interlocking cardinal teeth atop each valve. A distinct lunule lies ahead of the umbo and is arrowhead-shaped when valves are together. Quahogs differ from false quahogs (p. 60) in having low, rounded teeth on their inside bottom edge.

Southern quahogs *(Mercenaria campechiensis)* have their mid-shell growth lines clearly visible, the largest of which are as wide as a pencil lead. They are gray outside with occasional purple zigzags and broad rays. Inside they are mostly white but may have hints of purple.

Northern quahogs *(Mercenaria mercenaria)* look similar to their southern cousins, but they differ in having finer growth lines that are smooth in the center of larger clams. Their inner margin tends to be deep purple.

HABITAT: These clams live in the muddy sands of shallow bays and lagoons.

DID YOU KNOW? Quahog is pronounced KO-hog. Its genus *Mercenaria* translates to "payment," a reference to the wampum *(wampumpeg,* Algonquin for valuable string of beads) made from the quahog's purple parts. Surf-smoothed bits of this purple shell (middle image) stand out in the shell hash.

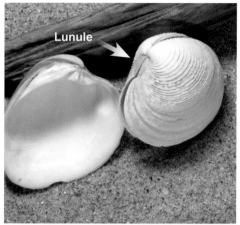
Southern quahog, max 5.9 in (15 cm)

Northern quahog, max 4.3 in (11 cm); and surf bits

Young southern (A) and northern (B) quahogs

59

Lady-in-waiting venus clam, max 1.6 in (4.1 cm)

Cross-barred venus clam, max 1.3 in (3.3 cm)

False quahog, max 1.5 in (4 cm)

Gray pygmy-venus clam, max 0.4 in (1 cm)

Venus Clams *(Lady-in-waiting, Cross-barred, False Quahog, and Gray Pygmy-)*

Lady-in-waiting, Cross-barred, and Gray pygmy- | False Quahog

IDENTIFYING FEATURES: Venus clams have three interlocking cardinal teeth and a distinct lunule.

Lady-in-waiting venus clams (*Puberella intapurpurea*) have strong concentric ridges that are serrated on the lower hind end. Colors are cream, tan, or gray, often with brownish streaks.

Cross-barred venus clams *(Chione elevata)* have sharp, concentric ridges that cross radial riblets. Even beach-worn shells show a distinct cross-hatched look. Inside colors often include deep purple.

False quahogs *(Pitar morrhuanus)* look similar to quahogs (p. 59) but have thinner, rusty-gray shells, smooth concentric lines, no teeth on their lower edge, and are white inside.

Gray pygmy-venus clams *(Timoclea grus)* have ribs crossed by growth lines and are cream or gray, often with a purple-brown streak covering the hind end.

HABITAT: Nearshore sand and rubble.

DID YOU KNOW? Exposed clams risk predation. The rough shell sculpture of the cross-barred venus hinders its ability to dig into sediment, but keeps it in place once it is there. Because of this, it prefers quiet waters where severe wave erosion is minimal.

Venus Clams
(Calico, Sunray, and Imperial)

IDENTIFYING FEATURES:

Calico clams *(Macrocallista maculata)* have smooth, creamy shells with blurry brown rectangles and smudges.

Sunray venus clams *(Macrocallista nimbosa)* have smooth, elongate shells that are purplish-brown with darker, narrow rays streaking from the umbo. Beach-worn shells may be bone white.

Imperial venus clams *(Chione latilirata)* have their shells thickened by 5–9 concentric, chunky rolls. They are whitish, light gray, or mottled tan with a few blurry rays.

HABITAT: Calico and imperial venus clams live in sand off beaches out to moderate depths. Sunray venus clams live in the muddy sands of shallow bays but may inhabit protected shoals within inlets and just offshore.

DID YOU KNOW? The sunray venus clam is popular with hungry oyster-catchers, gulls, and other local chowder fans. The thickened shell rolls of the imperial venus add strength and may help this shallow burrowing clam avoid predation from shell-cracking predators. This clam is most common within the offshore beds of calico scallops.

Calico clam, max 3.5 in (8.9 cm)

Sunray venus clam, max 6 in (15.2 cm)

Imperial venus clam, max 1.4 in (3.6 cm)

61

Thin cyclinella, max 1 in (2.5 cm)

Disc dosinia, max 3 in (7.6 cm)

Elegant dosinia, max 3 in (7.6 cm)

Softshell clam max 6 in (15.2 cm)

Venus Clams *(Thin Cyclinella, Disc and Elegant Dosinias)* and **Softshell Clam**

Thin Cyclinella and Elegant and Disk Dosinias

Softshell Clam

RELATIVES: Venus clams are in the family Veneridae. Softshell clams (Myidae) are more closely related to angelwings and piddocks.

IDENTIFYING FEATURES:

Thin cyclinellas *(Cyclinella tenuis)* are flat white with fine but irregular growth lines. They are smaller and have thinner shells than the dosinias.

Disc dosinias *(Dosinia discus)* have ivory, circular shells with sharp forward-pointing umbones. They have fine concentric ridges too narrow for most folks to count without a hand lens.

Elegant dosinias *(Dosinia elegans)* look similar to disc dosinias, but their flattened concentric ridges are broad, easily seen, and readily felt.

Softshell clams *(Mya arenaria)* are chalky and lumpy with fine concentric lines and an interior spoon-shaped projection from the hinge area.

HABITAT: Each of these clams lives in sand or muddy sand from outside the surf zone out to moderate depths.

DID YOU KNOW? Dosinias have a hinge ligament strong enough to keep their valves attached long after their demise, surf tumble, and beaching. Softshell clams, also known as steamer clams, are the target of a major fishery in the northeast states.

False Angelwing, Mud-piddock, and **Geoduck**

False Angelwing and Mud-piddock   *Geoduck*

False angelwing, max 2 in (5 cm)

RELATIVES: False angelwings (family Petricolidae) are closer kin to venus clams than to mud-piddocks (Pholadidae, the angelwing family) and geoducks (Hiatellidae), which are both distantly related to shipworms.

IDENTIFYING FEATURES:

False angelwings *(Petricolaria pholadiformis)* have winglike shells with a simple hinge margin bearing 3 (left valve) or 2 (right valve) cardinal teeth.

Atlantic mud-piddocks *(Barnea truncata)* are stubby, with pronounced shell gapes both front and rear. Freshly stranded shells have a long, delicate, spoon-shaped projection (apophysis) under the umbo.

Atlantic mud-piddock, max 2.8 in (7.1 cm)

Atlantic geoducks *(Panopea bitruncata)* are beached as large, off-white, lumpy, oblong shells that clearly did not close without large gapes at either end.

HABITAT: False angelwings and mud-piddocks bore into muddy clay, peat, or rotten wood on estuarine bottoms. Geoducks live in offshore burrows.

DID YOU KNOW? The siphons of an Atlantic mud-piddock extend 12 times its shell length. The Atlantic geoduck (pronounced GOO-ee-duk) lives in offshore burrows more than 5 ft (1.5 m) deep, which is why their beached shells are so uncommon.

Atlantic geoduck, max 9 in (23 cm)

Angelwing, max 6.7 in (17 cm)

Campeche angelwing, max 5 in (12.7 cm)

Campeche angelwing (L) and angelwing (R)

Angelwing
and Campeche Angelwing

Angelwing   *Campeche Angelwing*

RELATIVES: Angelwings (Pholadidae) are distantly related to shipworms.

IDENTIFYING FEATURES: Fragile, whitish, winglike shells with radial ribs. Freshly stranded shells have a long, delicate, spoon-shaped apophysis under the umbo.

Angelwings *(Cyrtopleura costata)* have a flared shell margin near the hinge that curves out at the umbo.

Campeche angelwings *(Pholas campechiensis)* have a flared margin in front of the hinge that curves out to cover the umbo. This membranous shell over the umbo is divided into several delicate compartments.

HABITAT: Angelwings bore into muddy clay, peat, or rotten wood on the bottoms of open bays. Campeche angelwings are most common in offshore clays. Abundant angelwings on the beach means that the beach has eroded into former estuary.

DID YOU KNOW? These angelwings live with much of their soft parts outside their shells, which are more important as a tool for burrowing than for protection. The shell's roughness abraids the sides of the angelwing's burrow allowing it to maintain a clear passage through which it moves quickly up and down. The angelwing's apophysis serves as an attachment site for its large foot muscles.

Piddocks and Shipworms

Wedge and Oyster Piddocks

Shipworms

Wedge piddock, max 2 in (5 cm), in wood

RELATIVES: Wedge and oyster piddocks share the family Pholadidae with true angelwings and are distantly related to shipworms (Teredinidae), which are bivalves, not worms.

IDENTIFYING FEATURES:

Wedge piddocks *(Martesia cuneiformis)* have wedge-shaped shells with rasplike concentric ridges and a groove from the umbo down. They burrow into wood.

Oyster piddocks *(Diplothyra smithii)* have weaker ridges than wedge piddocks and bore into thick shell and rock.

Oyster piddock, max 0.7 in (1.7 cm), in limestone

Shipworms (family Teredinidae) are evident as snaking **tunnels** through beached driftwood. The tunnels of these bivalves are lined with white, fragile, shell material. The animal is worm-shaped, having small, winglike, shell valves (generally deep in wood and not easily seen) in front and paddlelike "pallets" in the rear. *Teredo* shipworms have pallets hollowed like a vase, whereas *Bankia* has pallets of about 16 stacked funnels.

Shipworm tunnel shell, max 0.4 in (1 cm) diameter

HABITAT: Wedge piddocks and shipworms burrow in wood; oyster piddocks burrow in rock or shell.

DID YOU KNOW? Shipworms tunnel for protection and feed by filtering outside water taken in by siphon tubes. Their shell valves close like jaws to grind wood and their rear pallets plug their tunnel to prevent dehydration.

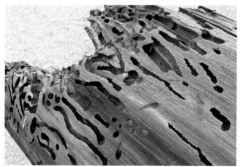

Shipworm tunnnels in driftwood

65

Circular Chinese-hat

Adam's Miniature Cerith

Amethyst gemclam

Contracted corbula

Pointed nutclam

Atlantic nutclam

Atlantic abra

Lunate crassinella

White strigilla

Many-line lucine

Miniature Mollusks

RELATIVES: Gastropods and bivalves

IDENTIFYING FEATURES: These shells are too small to be seen by folks on a casual stroll and are all less than about 0.25 inch (8 mm) as adults. The path into the amazing world of itty-bitty shells is traveled by those on their hands and knees. Peering into drift piles at the recent strand line will reveal many of this page's petite species in addition to miniature versions of the larger species shown on previous pages.

Dinky gastropods:

Circular Chinese-hat *(Calyptraea centralis),* family Calyptraeidae, rare north of Hatteras

Adam's Miniature cerith *(Cerithium muscarum),* family Cerithiopsidae, shell rubble, all beaches

Itty-bitty bivalves:

Amethyst gemclam *(Gemma gemma),* family Veneridae, all beaches

Contracted corbula *(Caryocorbula contracta),* family Corbulidae, near inlets, all beaches

Pointed Nutclam *(Nuculana acuta),* family Nuculidae, all beaches

Atlantic nutclam *(Nucula proxima),* family Nuculidae, all beaches

Atlantic abra *(Abra aequalis),* family Semelidae, all beaches

Lunate crassinella *(Crassinella lunulata),* family Crassatellidae, all beaches

White strigilla *(Strigilla mirabilis),* family Tellinidae, rare north of Hatteras

Many-line lucine *(Parvilucina multilineata),* family Lucinidae, rare north of Hatteras

Fossil Shells

Giant Oyster

Other Fossils

Giant oysters, extinct cousins of modern kinds

WHAT ARE THEY? Fossil shells are mineralized remnants of mollusks that lived long ago. Although there is no age requirement, most shell fossils are pretty old and may be from species that are not around any more. Fossilized shells are chalky white or are stained by grayish or orange minerals. Some have become surrounded by limestone rock.

Extinct **giant oysters** *(Crassostrea gigantissima)* are gray, massively thick bivalve shells that wash ashore from ancient reefs off the beach. **Fossil venus clams,** larger than our modern quahog (p. 59), lived in estuaries where the beach is now. A variety of fossil shells are common wherever **coquina limestone** is found. Others are found where inlets cut through deposits and where inland fill has been dumped near the beach.

Fossil venus clams

SIZE: Most beach fossils are hand-sized or smaller. Giant oysters are up to 2 ft (60 cm).

ORIGIN: Because beach dynamics mix many ages of material together, shell fossils may span ages from recent times to many millions of years. Most beach fossils are from the Pleistocene Epoch (10,000 to 1.8 million years ago). During this time, our Southeastern coastline advanced and retreated with four major changes in sea level, each corresponding to an ice age. Some of our oldest fossils of animals that lived locally are

Fossilized bivalves from shelly coquina limestone

67

Fossil gastropods

giant oysters, which made up reefs off Topsail Island, NC, about 25 million years ago. However, ancient rivers washing material from the Piedmont has sprinkled beaches with some fossils as old as the Triassic Period, 200–250 million years ago. An assortment of these fossils include:

A. *Mitra (Pleioptygma) heilprini,* family Mitridae

B. *Vasum horridum,* family Vasidae

C. *Cancellaria plagiostoma,* family Cancellariidae

D. *Fusinus caloosaensis,* family Fasciolariidae

E. *Conus adversarius,* family Conidae

F. *Fasciolaria apicina,* family Fasciolariidae

G. *Crucibulum multilineatum,* family Calyptraeidae

H. *Hanetia mengeana,* family Muricidae

I. *Eupleura intermedia,* family Muricidae

The seashells above have been extinct for roughly two million years.

J. *Nodipecten* **sp.,** family Pectinidae, extinct 1–15 million years

K. *Nodipecten nodosus,* family Pectinidae, still with us (see page 42)

L. *Chione latilirata,* family Veneridae, still with us (see page 61).

M. *Eucrassatella speciosa,* family Crassatellidae, still with us.

N. *Chione* **sp.,** family Veneridae, still with us (see page 60).

O. *Plicatula gibbosa,* family Pectinidae, still with us (see page 44).

Fossil bivalves

Shell Color Variation

Although some shells are most colorful in life, other shells turn a variety of colors after their inhabitants die. These colors depend on the shell's afterlife experiences. Black shells were likely darkened by iron sulfide after burial in sulfurous muck. A beach speckled with numerous black shells shows that the surf zone was once a lagoon behind the barrier island. Pink, rust, or brown are the colors most shells turn after decades of exposure to the iron oxides in underwater beach sediments.

Colorful incongruous arks from the same beach

Shells out of water exposed to sunlight bleach pale gray or chalky white in just a few decades. Although glossy white shells are probably recent, bone-white shells may be fossils that have changed little over the millennia they've aged. Varied shell colors hint that most beaches in the Southeast have shells of mixed origin and age, but most shells on the upper beach are pretty old. There, the average shell is typically a few thousand years old.

The colors of recently living shells vary with what the mollusk ate and where it lived, in addition to what genes the animal inherited from its parents.

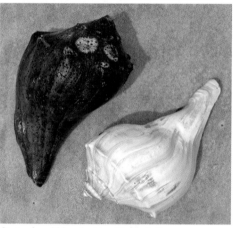

Stained and bleached knobbed whelks

Inherited color traits are dictated by use of four shell-color pigments—dark melanins, yellow carotenoids, green porphyrins, and indigoids that come in either blue or red. The iridescent sheen of nacre (mother of pearl) is not from pigment, but instead comes from the refraction of light by transparently thin layers of calcite. These rainbow colors are not seen while the mollusk is alive and do not contribute to its well-being. Nacre serves to smoothen the shell surface next to the mollusk's soft body.

The interior nacre of a penshell

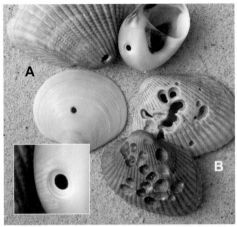

Boreholes from gastropods (A) and bivalves (B)

Boring sponge perforations in a quahog

Polydorid polychaete worm grooves

Shell Wars (Shell Bioerosion)

Beached mollusk shells often bear clues to how they met their demise and who made use of them after their death. This evidence includes boreholes, perforations, and grooves.

Shells with single, circular **boreholes** were likely eaten by a predatory gastropod. Atlantic oyster drills (p. 32) leave a straight hole, whereas thick-lipped drills leave a slightly beveled hole. Shark eye snails (pp. 14–15) leave a countersunk, circular borehole that has an outer diameter twice the inner diameter. Note in the top image that shark eyes can be cannibals. Two tactics used by hole-boring gastropods include edge drilling and umbo drilling. Drilling at the valve edge is the fastest because the shell there is thinner, but an edge-drilling snail risks a pinched proboscis from closing valves. Drilling the thick umbo is safer but takes time, during which a snail may have its prey stolen or become a meal itself. Drilling a thick shell can take a snail the better part of a day. Bivalve shells also may be penetrated by other bivalves like *Gastrochaena,* which leave oblong boreholes in either shell or rock. This bean-shaped clam lives out its life within the pit it forms.

Scattered **perforations** in a shell were likely made by boring sponges These sponges partially acid-digest living and dead shells and invade them as living space.

Other animals that use shells as living space include polychaete worms (bristle worms) like polydorids, which leave snaking **groove** marks. The worm makes these router-tool indentations by softening the shell with secreted acid and rasping with its bristled body.

Mollusk Bits and Pieces

For every whole shell found on a beach there are thousands of bits and pieces. Some surf-worn shards have the clear distinguishing features of the original shell. Do you recognize these?

A. Penshell (p. 41) inside nacre

B. Atlantic giant cockle (p. 50) ribs

C. Atlantic giant cockle (small) (p. 50) umbo and hinge area

D. Giant tun shell (p. 17) body whorl

E. Lettered olive (p. 33) body whorl

F. Northern quahog (p. 59) purple margin

G. Florida fighting conch (p. 9) body whorl

H. Clench helmet (p. 18) aperture teeth

I. Channeled whelk (p. 27) body whorl and spire

J. Shark eye (pp. 14–15) body whorl and spire

K. Campeche angelwing (p. 64) umbo end with apophysis

L. Lightning whelk (p. 26) left-handed columella

M. Knobbed whelk (p. 24) right-handed columella

N. Eastern oyster (p. 46) hinge scar

O. Channeled duck clam (p. 53) umbo and hinge

P. Atlantic figsnail (p. 17) siphon canal of body whorl

Q. Angelwing (p. 64) posterior margin

R. Scotch bonnet (p. 18) aperture shield and lip

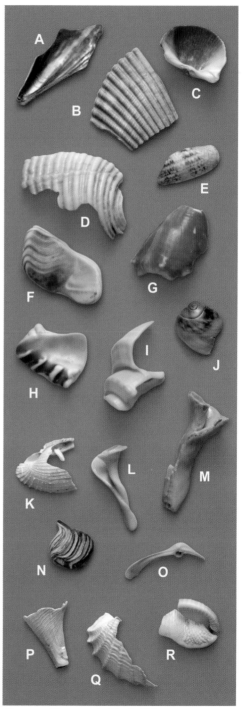

Some readily recognizable shell fragments

"It is perhaps more fortunate to have a taste for collecting shells than to have been born a millionaire."

~ from *Lay Morals,*
by Robert Louis Stevenson 1850–1894

Suggested Resources

Hugh J. Porter, Lynn Houser. *Seashells of North Carolina.* Raleigh, NC: North Carolina Sea Grant, 2000.

Lee, Harry G. *Marine Shells of Northeast Florida.* Jacksonville, FL: Jacksonville Shell Club, 2009.

Morris, Percy A. *A Field Guide to Shells of the Atlantic and Gulf Coasts and the West Indies.* Boston, MA and New York, NY: Houghton Mifflin Company, 1973.

www.jaxshells.org [for updated seashell identification and nomenclature]

http://www.saltwatertides.com [for tidal times and amplitudes]

Index

Entries in **bold** indicate photos and illustrations.

Here are some other books from Pineapple Press on related topics. For a complete catalog, visit our website at www.pineapplepress.com. Or write to Pineapple Press, P.O. Box 3889, Sarasota, Florida 34230-3889, or call (800) 746-3275.

Living Beaches of Georgia and the Carolinas by Blair and Dawn Witherington. A guidebook to the over 600 miles of wave-swept coastlines of Georgia and the Carolinas. Accurate identification information for birds, shells, plants, and animals. Over 1000 photos. (pb)

Florida's Seashells by Blair and Dawn Witherington. Accounts, maps, and color photos for over 250 species of mollusk shells found on Florida's beaches. (pb)

Florida's Living Beaches: A Guide for the Curious Beachcomber by Blair and Dawn Witherington. A guide to the myriads of plants, animals, minerals, and manmade objects you can find along Florida's coastline. Descriptive accounts of 822 items, 983 color images, and 431 maps. (pb)

Coastal South Carolina by Terrance Zepke. Includes brief histories and fast facts for the islands and communities, maps, and tourism resources. (pb)

Coastal North Carolina by Terrance Zepke. Find quick histories of islands, towns, and regions; main sites and attractions; opportunities for recreation; and festivals and events. (pb)

Ghosts of the Carolina Coasts by Terrance Zepke. These 32 spine-tingling ghost stories take place in prominent historic structures of the region. (pb)

Ghosts and Legends of the Carolina Coasts by Terrance Zepke. More spine-chilling tales and fascinating legends from the coastal regions of North and South Carolina. (pb)

Ghosts of the Georgia Coasts by Don Farrant. Meet the ghosts that haunt Georgia's historic places, including crumbling slave cabins, plantation homes, ancient forts. (pb)

Best Ghost Tales of North Carolina, Second Edition, and *Best Ghost Tales of South Carolina* by Terrance Zepke. The actors of the Carolinas' past linger among the living in these tales. (pb)

Pirates of the Carolinas, Second Edition, by Terrance Zepke. Thirteen of the most fascinating buccaneers in the history of piracy, including Blackbeard, Anne Bonny, and Calico Jack. (pb)

Lighthouses of the Carolinas by Terrance Zepke. Eighteen lighthouses aid mariners traveling the coasts of North and South Carolina. Here is the story of each, along with visiting information and photographs. (pb)

Georgia's Lighthouses and Historic Coastal Sites by Kevin M. McCarthy. Illustrations by William L. Trotter. The lighthouses, forts, historic homes, plantations, and churches of Georgia's coast. (pb)